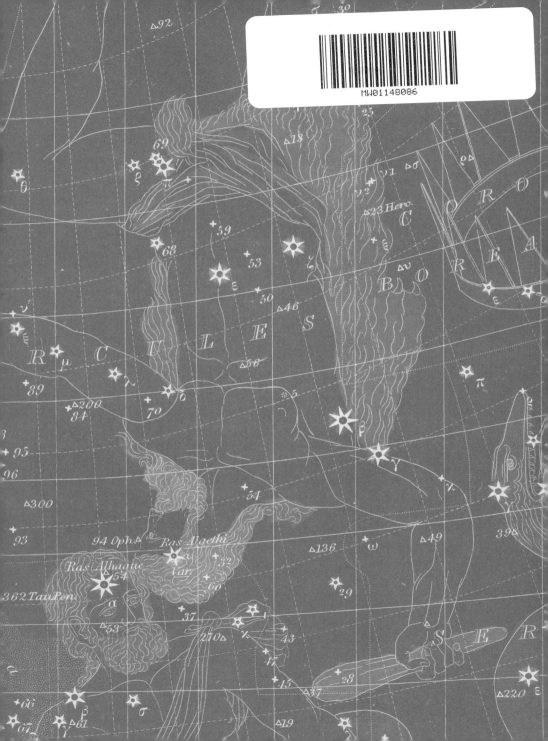

# FIRST CONTACT

## The Story of Our Obsession with Aliens

BECKY FERREIRA

WORKMAN PUBLISHING • NEW YORK

Text Copyright © 2025 by Becky Ferreira

Hachette Book Group supports the right to free expression and the value of copyright. The purpose of copyright is to encourage writers and artists to produce the creative works that enrich our culture.

The scanning, uploading, and distribution of this book without permission is a theft of the author's intellectual property. If you would like permission to use material from the book (other than for review purposes), please contact permissions@hbgusa.com. Thank you for your support of the author's rights.

WORKMAN
Workman Publishing
Hachette Book Group, Inc.
1290 Avenue of the Americas
New York, NY 10104
workman.com

Workman is an imprint of Workman Publishing, a division of Hachette Book Group, Inc. The Workman name and logo are registered trademarks of Hachette Book Group, Inc.

Cover design by Jack Dunnington
Design by Raphael Geroni

All photo and art copyright/credits information can be found on page 263

The publisher is not responsible for websites (or their content) that are not owned by the publisher.

Workman books may be purchased in bulk for business, educational, or promotional use. For information, please contact your local bookseller or the Hachette Book Group Special Markets Department at special.markets@hbgusa.com.

Library of Congress Cataloging-in-Publication Data is available.

ISBN: 978-1-5235-2775-5

First Edition September 2025

Printed in Shenzhen, China (APO), on responsibly sourced paper.

10 9 8 7 6 5 4 3 2 1

To Mum and Dad,
thanks for all the fish

# TABLE OF CONTENTS

### PROLOGUE
**vi — THE GREAT SEARCH**
*Humans have suspected for tens of thousands of years that we are not alone in the universe.*

### CHAPTER ONE
**1 — ANCIENT ALIENS**
*The debate over the existence of extraterrestrial life takes many twists and turns throughout history.*

### CHAPTER TWO
**29 — WHERE IS EVERYBODY?**
*Astronomical advances in the twentieth century unveil an unexpectedly silent universe.*

### CHAPTER THREE
**55 — POP CULTURE ALIENS**
*We haven't found real aliens yet, but our countless fictional versions will tide us over until we do.*

### CHAPTER FOUR
**83 — FLYING SAUCERS**
*As scientists look to find life in space, many people believe it is already here on Earth.*

## CHAPTER FIVE

**121 THE ALIEN NEXT DOOR**

*The search for life in our solar system includes many tantalizing nearby worlds.*

## CHAPTER SIX

**155 DEEP SPACE**

*The search for life beyond the solar system has been supercharged by exoplanet discoveries and hunts for technosignatures.*

## CHAPTER SEVEN

**191 CONTACT**

*Preparations for a post-detection future are in full swing.*

## CHAPTER EIGHT

**217 NO CONTACT**

*What would it mean to humanity if we never found alien life?*

## CHAPTER NINE

**233 WHEN WE BECOME THEM**

*If humans become a spacefaring diaspora, we may become the extraterrestrials we seek.*

## AFTERWORD

**248 TERRESTRIAL INTELLIGENCE**

250 Sources
256 Index
261 Acknowledgments
263 Photo Credits
264 About the Author

## PROLOGUE

# The Great Search

SUPERNOVAS. ASTEROID STRIKES. GALACTIC JETS that incinerate anything, or anyone, in their path. Collapsed magnetic fields. Atmospheres blown off into space like dandelion seeds. Atmospheres that curdle and smother. Worlds hurled from their stars. Worlds hurled *into* their stars.

Let's be real: The universe seems like a psychopath that wants us dead. Earth is an outlier, a middle finger raised against an inhospitable expanse. We are so very lucky to be alive. It sure doesn't look like much else is around these parts. But that hasn't stopped humans from ardently, defiantly believing that we are not alone in the universe.

For tens of thousands of years, our species has suspected that otherworldly beings inhabit the skies—and that some might even walk among us here on Earth. It's as if some innate part of us has always yearned to know: If we are here, then who is out there?

We still don't know the underlying truth behind this ancient premonition, but we are in the midst of an epic quest to confirm our hunch. This book traces the lineage of that effort back to its deep roots in prehistory. It will size up the massive footprint that aliens have left in our imaginations. It will also catalog the myriad clever techniques that we have designed to search for life beyond Earth. It will envision the fateful moment when we discover something—a fossil, a message, a fleeting hint of metabolic activity—that at last validates our ancestral belief that there are *others*. It will also acknowledge that many people believe this moment has already happened.

Given that nobody really knows if aliens exist, it is striking to consider that so much thought, time, and resources have been invested into the hypothetical. Yet the prospect of finding another form of life—be it extant or extinct, simple or complex, nearby or across the universe—is simply too tantalizing to ignore. This hunt for extraterrestrial life transcends eras and borders, spurring people to work together across time and space toward a common goal that may not ever materialize.

This great search will end, someday, one way or another. We will either find life elsewhere, filling in that most irresistible datapoint of our universe—or we will have to come to terms with our fate as a biological fluke in an apparently barren universe. But our quest to find aliens is, at heart, an experiment in self-examination. No matter how far we peer into space and no matter how many apparitions we conjure, the lens will always turn back on ourselves.

||||||||||||||||||||||||||||||||||||||||||||||||||||||||||||

Heaven and Earth are large, yet in the whole of empty space they are but as a small grain of rice. Empty space is like a kingdom and heaven and Earth no more than a single individual person in that kingdom. How unreasonable it would be to suppose that, besides heaven and Earth, we can see there are no other heavens and no other Earths?

—TANG MU, *thirteenth-century philosopher*

||||||||||||||||||||||||||||||||||||||||||||||||||||||||||||

# CHAPTER ONE

# ANCIENT ALIENS

# We have met others before.

**LONG AGO, IN A LOST WORLD OF MAMMOTHS AND DIRE** wolves, we crossed paths with travelers who were so much like us that we made children together. Our relationship with these others, the Neanderthals and Denisovans, is shrouded in mystery, but it's clear we met again and again, across continents and millennia, until there was nothing left of them but bones, artifacts, and the genetic signatures that we bear in our cells today as the last humans on Earth.

Then there were the wild ones, those wolfy ancestors of dogs, that we admired from afar. We made contact, struck a deal, and they left a doomed lupine line for a future by our side. They were the first of many animals brought into our fold for their strength, flesh, and companionship. Even today, we are still trying to open communications with fellow Earthlings, from dolphins to octopuses, birds to insects, and our own grand family of primates.

Given the dazzling variety of intelligences around us, it's no surprise that somewhere in the shadows of prehistory, our ancestors began to look skyward in search of others. As far as we know, they never met anyone from "out there," and we have fared no better today in spite of all our advances.

But though we are separated from these ancients by time, we share their instinct to personify the skies, imbuing its features with familiar names, narratives, and prophecies. For as long as we've been recognizably human, space has reflected our dreams and prejudices. That instinct to believe that there is something alive, perhaps sentient, beyond Earth continues to this day in the search for extraterrestrial life.

Chapter One: ANCIENT ALIENS ▶ 003

# Sky Lore

**W**E KNOW LITTLE ABOUT WHEN, OR WHY, humans shifted from an ambient awareness of the sky to recognizing specific celestial objects and patterns. But it happened a very long time ago—tens of thousands of years into the past, at minimum.

Early skywatchers may not have perceived clear distinctions between local earthly phenomena, such as auroras and thunder, and the much more distant realm of the planets and stars. But they did see a great celestial tapestry that contained elements of order. They recognized the repeating cycles of the Sun and Moon, punctuated with capricious surprises—shooting stars, comets, and the glare of stars that exploded far from Earth, visible even in daytime.

▲ Hipparchus inspects ancient Greek skies. Let's bring back stargazing in togas.

Somewhere along the way our ancestors learned to use the sky as a guide, a clock, a starry storyscape that connected cycles of life and time. Tales were shared by firelight, in the chambers of caves, on foot to the next horizon—tales about how those beacons above signaled the existence of others, elsewhere, off Earth.

The majority of human star lore has long since faded from memory, but the surviving tales continue to shape our current expectations about extraterrestrial life. Ancient myths about the cosmos are the imaginative nebular clouds from which the modern concept of aliens condensed. And yet our advances in science and technology have not freed us from our familiar narrative ruts and well-worn tropes, including:

- **MESSAGES FROM ON HIGH:** Today, we scan the sky with sophisticated telescopes, looking for heavenly signals of intelligence and other clues about life beyond Earth. Ancient peoples, equipped with only the naked eye, pioneered this belief that the sky is a canvas for communications, and they perceived messages written in the language of comets, eclipses, and other cosmic phenomena.

- **SUPERIOR BEINGS AND TECHNOLOGIES:** Countless myths envision powerful beings that inhabit the sky, including divinities that deliver advanced technologies to Earth. Chief among them is the Greek god Prometheus, who stole fire from his fellow deities and gave it to humanity. The expectation of finding more advanced aliens and benefitting from their knowledge and industries remains an undercurrent in the modern search for extraterrestrial life.

- **COSMIC KINSHIP:** Many cultures have imagined themselves as the descendants of celestial objects, or believed that they will ascend to the skies to join their ancestors after death. The sense that we share an elemental kinship with the cosmos has turned out to be literally true, as the materials in our bodies were forged by stars. The astronomer Carl Sagan famously drew on this scientific revelation and its mythological precursors when he said: "The cosmos is within us. We are made of star-stuff."

::: **VISITATION AND ABDUCTION:** Celestial divinities in tales told across the ages have enjoyed dropping in on Earth to observe people, often in disguise. Sometimes, they'll go so far as to kidnap someone, possess their body, or impregnate them (looking at you, Zeus). Modern science fiction is packed with similar stories about alien visitation, abduction, and possession; in addition, thousands of people have reported that they have actually experienced these extraterrestrial encounters.

▲ This scene combines many common abduction elements: a flying saucer, a tractor beam, and, yes, implied experimentation on livestock.

## "THE LOST PLEIAD"

The Pleiades star cluster is bright, distinctive, and visible from both hemispheres, distinguishing it as one of the most recognizable and culturally important patterns in the night sky. Countless stories have been told about this cluster, many of which cast the most radiant stars as a group of seven young women, often sisters, who are fleeing the rapacious advances of a male hunter, or group of hunters, represented by the nearby constellation Orion.

What's weird about the cross-cultural resonance of the Seven Sisters story is that there are only six distinctive stars in the modern constellation. So what gives? Where is the missing Pleiad?

Astronomers Ray and Barnaby Norris believe this question is a key that could unlock the oldest-known story shared by humans, a yarn of yore that dates back 100,000 years to a time before our species dispersed beyond Africa.

In this long-lost age, our ancestors would have seen two stars in the cluster, Atlas and Pleione, as distinct sparkles of light. Since then, those stars have moved so close together from our perspective on Earth that they appear as one to the naked eye. This gradual transformation of the cluster could explain the prevalence of stories about a "lost Pleiad"—an early extraterrestrial personality—that crop up in European, African, Asian, Indonesian, Indigenous American, and Aboriginal Australian cultures.

Of course, it's hard to imagine how we might verify this hypothesis, barring the invention of time travel and/or *Dune*-style genetic memory. It's also worth noting that this star cluster inspired many stories that don't fit the "lost Pleiad" mold; my personal favorite comes from the Mono Indigenous Americans of the Sierra Nevada, who say the cluster is a group of wives who abandoned their husbands so that they could eat onions eternally without spousal complaints about the stink.

---

▶ In addition to inspiring amazing stories, the Pleiades have served as markers of seasonal change for tens of thousands of years.

Still, it's tantalizing to imagine that we might trace the lineage of this story all the way back to those ancient skywatchers, who had yet to set out across continents and oceans. Though we will likely never know whether the basic story dates back that far, there is no doubt that humans seem to almost instinctively picture personalities in the sky, whether they are liberated onion-breathed women or exotic alien life-forms.

> **ALIEN** is derived from the Latin *alius*, meaning "other."

## COSMIC PLURALISM AND THE PROTO-ALIEN

About 2,600 years ago, a bunch of weirdos who lived on the shores of the Mediterranean Sea began to speculate that the natural world could be explained without invoking supernatural entities. This protoscientific idea stood in contrast to popular myths that ascribed natural events to the whims of a highly entertaining cast of divine characters prone to cavorting, battling, and occasionally eating their own babies (hiya, Cronus).

The paradigm shift gave rise to the earliest known iterations of cosmic pluralism: the idea that there are many real physical worlds, and perhaps even life-forms, beyond Earth. It's a conclusion that might seem trivial today, as rovers scramble across Mars and telescopes spot exoplanets by the thousands. But before the advent of modern astronomy, people were vexed, awestruck, and deeply opinionated about the basic nature of objects in the sky. There were once real stakes to the belief in other alien worlds—not just figurative stakes, but the kind on which you might conceivably get burned to death (see "Giordano Bruno," page 18).

## ANAXAGORAS OF CLAZOMENAE

Born around 500 BCE, Anaxagoras of Clazomenae grew up on the scenic Anatolian coast of what is now Turkey. After moving to Athens as a young man, Anaxagoras became an influential philosopher and an early proponent of the idea that celestial bodies, such as the Moon, were natural environments that might host life-forms of their own.

By viewing celestial objects as tangible entities, Anaxagoras was able to make many major breakthroughs—for instance, he was the first person known to explain the true mechanics of eclipses. The authorities were decidedly unchill about all this philosophizing, however, and Anaxagoras was condemned to death for impiety—a charge based on his rejection of religious teachings—after he kept insisting the Sun was a hot and large natural object, not some hunky god driving a chariot across the sky. He was spared thanks to his friendship with Pericles, the great Athenian statesman, who is said to have persuaded the accusers to downgrade the sentence to exile.

Anaxagoras's story represents a major milestone in the history of extraterrestrials because he envisioned life as a natural process that could emerge anywhere with the right ingredients and circumstances. But his life also foreshadowed a major tension that would frequently erupt between a belief in alien worlds and the dominant dogmas of the times.

▲ Anaxagoras, the original eclipse chaser.

# Lucian's *True History*
## (aka, Don't F*ck the Plant People)

**LUCIAN OF SAMOSATA,** a prolific Syrian writer in the second century, is considered a pioneer of alien science fiction because of his cheekily named satire *True History* (which kicks off with the disclaimer that the whole thing is a total fabrication). The work parodies classic epics by Homer and others by recounting a series of completely over-the-top tall tales about a lost ship crew that gets transported to the Moon, where they encounter strange alien life-forms enmeshed in a battle with the Sun.

Lucian's freewheeling satire also has the honor of pioneering a trope that has become ubiquitous in modern science fiction: human-alien copulation. During their voyages in strange lands, the crew meet a weird species of botanical tree-women and, naturally, some of the sailors simply must have sex with them.

"Two of my companions who got involved with them could not manage to get free, but became attached to them by their private parts," Lucian's narrator laments. "As this happened, their fingers grew into small shoots and the tendrils twisted about them so tightly that they were about to bear fruit. We abandoned our companions and fled to our ship."

Behold, a lesson from antiquity that still rings across the ages: Beware of alien hookups.

▲ It's so hard to find a good alien-woman who is not part tree these days.

# Aliens and Religions

> "Space contains such a huge supply of atoms that all eternity would not be enough time to count them and the force which drives the atoms into various places just as they have been driven together in this world. So we must realize that there are other worlds in other parts of the universe, with races of different men and different animals."
>
> —TITUS LUCRETIUS CARUS, *Roman philosopher*
> *(first century BCE)*

GRECO-ROMAN PHILOSOPHERS CONTINUED TO promote theories about extraterrestrial worlds and lifeforms for many centuries after Anaxagoras's life and death. This pro-alien school of thought was linked to atomism, the theory that all matter, from humans to planets, arose out of tiny parts called *atoms* from across the universe. In other words, the same events that gave rise to life on Earth are occurring throughout the cosmos, suggesting that life is likely to exist on other worlds.

> **EXOTHEOLOGY:** A field of scholarship that examines how theological issues apply to extraterrestrial life and intelligence, also known as astrotheology.

Eventually, atomism was subducted under the rising philosophies of Plato and Aristotle, who held that Earth is a unique and central entity in the universe and that outer space is a whole other ball game governed by completely different rules. The concept of worlds similar to Earth, and likewise inhabited by living creatures, was disfavored by this view.

Early Christian philosophers adopted the Platonic centrality of humans in their conception of God's universe, leading to a hiatus of cosmic pluralism within the Church's sphere of influence for more than 1,000 years. Even so, Christian cosmology is packed with examples of otherworldly entities, such as angels or the Holy Ghost, visiting our earthly realm to interact with humans and influence events. In this way, Christian myths are built upon many of the same popular narratives that gave rise to modern alien lore, even though early Christians believed that Earth was the only inhabited world in the cosmos.

While Christianity initially shelved the possibility of aliens in a literal sense, many faith traditions naturally gel with the assumption that humans are not alone in the universe. Hinduism, a religion that is about twice as old as Christianity, is built around the concept of infinite realms populated by infinite different souls and beings, a worldview that not only allows, but arguably predicts, the existence of alien life.

As Islam flourished around Mecca in the 600s CE, some early Muslim scholars argued that the dominion of Allah also extended to other planets. Indeed, Allah is frequently described as the "Lord of the Worlds"—as in plural *worlds*—in the Quran. Even today, many Muslims interpret this scriptural phrasing as proof that extraterrestrial life is compatible with Islam.

## Chapter One: ANCIENT ALIENS ▶ 013

> "Maybe you see that Allah created only this single world and that Allah did not create humans besides you. Well, I swear by Allah that Allah created thousands and thousands of worlds and thousands and thousands of humankind."
>
> —IMAM MUHAMMAD AL-BAQIR, *a major leader of Shia Islam who lived from 676 to 732 CE*

In the Middle Ages, debates broke out within Judaism about the centrality of humans, and Earth, in the universe. The twelfth-century rabbi Moses ben Maimon, better known as Maimonides, warned in his masterpiece *The Guide for the Perplexed* that "it is of great advantage that man should know his station, and not erroneously imagine that the whole universe exists only for him." Maimonides believed that there might be other children of God elsewhere in the universe, and that their existence would not diminish God's love for his children on Earth.

But perhaps the most enduring openness to aliens comes from the countless folk and Indigenous traditions that have been practiced since time immemorial, often in the shadows and margins of giant organized religions. Many of these faith systems assume that life flows through all kinds of natural objects, like the stars or rivers, and that there is no terrestrial limit to these animating forces.

▲ The great philosopher and astronomer Maimoindes, pictured here with resting genius face.

## ⋮⋮⋮ An Alien Princess Folk Tale

**ONE DAY LONG** ago, a bamboo cutter was out in the forest when he spotted a mysterious glowing stalk. After slicing it open, he discovered a thumb-sized infant curled up inside. Overjoyed, the cutter and his wife adopted the baby girl and named her Kaguya-hime ("Shining Princess").

The proud parents marveled as their daughter rapidly transformed into a majestic woman over the course of a few short months. Kaguya-hime attracted many suitors, including the emperor of the realm, but she rejected their advances, one by one. Eventually, she fell into a depression and spent night after night gazing listlessly at the Moon.

One evening, she revealed to her parents that the Moon was her true home and that she had lived there as a princess before she was reborn in her earthly bamboo stalk. Moreover, the time had come for her return to her lunar kingdom, leaving her parents with only gifts, memories, and heartbreak.

This deeply enchanting story, known as "The Tale of the Bamboo Cutter," has been told in Japan for more than 1,000 years. It is one of the most ancient and beloved examples of monogatari, a tradition of prose fiction similar to epics. It's also essentially a story about alien visitation.

While Kaguya-hime is depicted in human form, she has otherworldly qualities that ultimately make her a bad fit for Earth. And though the raconteurs who first spread this story didn't share our understanding of alien life, the tale has inspired many modern retellings with imaginative extraterrestrial settings (including the manga sensation *Sailor Moon*).

▲ Princess Kayuga-hime ascends back to the Moon in woodblock print from 1891.

# The Copernican Revolution

**T**HE BELIEF THAT EARTH SITS AT THE CENTER of the universe and that the Sun and all other worlds orbit around it is often derided as one of humanity's most epic brain farts (a category for which there has long been fierce competition). But Earth-centric cosmologies, like the one put forth by the polymath Claudius Ptolemy, were often ingenious and practical for predicting astronomical cycles, including eclipses, in spite of their flawed foundations. This helps to explain why the Ptolemaic model and its variants remained so influential well into the Middle Ages, more than a millennium after they were conceived.

Still, many thinkers over the centuries chafed under the shortcomings of the theory. You can almost sense the exasperation of Ibn al-Haytham, a tenth-century polymath who lived in what is now Iraq, in his gripe that "Ptolemy assumed an arrangement that cannot exist, and the fact that this arrangement produces in his imagination the motions that belong to the planets does not free him from the error..." That's actually a pretty sick burn for the tenth century.

> **ARISTARCHUS OF SAMOS**, a Greek astronomer who lived in the third century BCE, presented the first known model of a Sun-centric (aka heliocentric) universe.

Nicolaus Copernicus, born in 1473 in what is now Poland, at last relieved these astronomical headaches with his landmark work *Six Books on the Revolutions of the Heavenly Spheres,* which placed the Sun in its rightful place at the center of the solar system. Though Copernicus was not the first to conceive of a heliocentric model, his meticulous work propelled its popularization across Europe. The notion that Earth was just one of many planets orbiting the Sun raised the question of whether other stars might host their own systems of worlds, reviving debates about cosmic pluralism.

> **COPERNICAN PRINCIPLE:** The insight that Earth and humans do not occupy a special or privileged place in space, and therefore observations from Earth represent an average view of the universe.

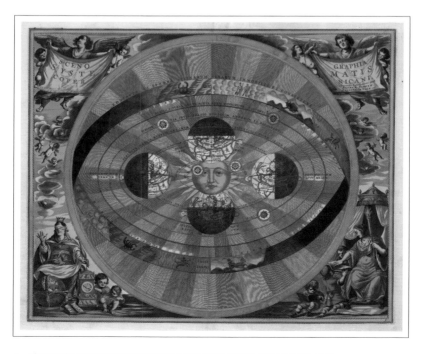

▲ The Sun looking smug in its rightful place at the center of the solar system.

## The Moons of Jupiter

▲ Io   ▲ Europa   ▲ Ganymede   ▲ Callisto

**THE GREAT ASTRONOMER** Galileo Galilei knew Copernicus was right: Earth orbited the Sun. But the Catholic Church was executing Copernican thinkers for espousing such heresy. What to do?

This question gained new urgency during the brisk winter nights of 1609 to 1610, which Galileo spent peering through a homemade telescope at a handful of "stars" that seemed to weirdly hover around the planet Jupiter. Eventually, he realized that these mysterious lights were actually satellites orbiting the gas giant, which are now known as the Galilean moons: Io, Europa, Ganymede, and Callisto.

This major discovery not only corroborated the Copernican model of a Sun-centric solar system; it also revealed that planets could have orbiters of their own. Galileo shared this revelation in his March 1610 treatise, *Sidereus nuncius (The Starry Messenger)*.

"These four little moons exist for Jupiter, not for us," wrote Johannes Kepler in response to *The Starry Messenger*. "Each planet in turn, together with its occupants, is served by its own satellites. From this line of reasoning we deduce with the highest degree of probability that Jupiter is inhabited."

Galileo's obstinate support of the Copernican model was enough to get him on the wrong side of the Catholic Church, even without any speculations about extraterrestrial life. He would spend the last decade of his life under house arrest. But the sentence could not brook the full-on flood of alien speculation that was about to sweep across seventeenth-century Europe, pushing traditional dogmas to the brink.

## Giordano Bruno: Visionary Alien Shit-Disturber

**GIORDANO BRUNO**, a spectacularly cantankerous philosopher born in 1548, walked into history like a wrecking ball and left it in a literal blaze of righteous fury that remains legendary to this very day.

Bruno had a superhuman imagination and a flagrant contempt for authority, qualities that were uniquely ill-fitted to the repressive era into which he was born. But while he paid dearly for his boldness, Bruno more than any other figure is responsible for igniting the modern obsession with extraterrestrial worlds and the life-forms that may inhabit them.

It wasn't that he even wrote about aliens all that much, but that he showed no fear about openly promoting cosmic pluralism, the concept that many worlds exist beyond Earth, and its implications for life on Earth and beyond it. All this came at a time when the Catholic Church was becoming particularly un-copacetic with that sort of talk. Though some Catholic theologians had speculated about aliens with no pushback from the Church, including the 15th-century German bishop Nicholas of Cusa, Bruno ignited pushback as the Church confronted new contradictions between its dogmas and scientific discoveries.

Bruno came of age alongside the Copernican model, which was published just a few years before his birth in Nola, a commune outside of Naples. The Church was loath to leave the warm bath of geocentrism, where humanity occupied a special earthly place under God's watch. It had, after all, just spent the past millennium and change building a theological fortress around this Earth-centric model, one that strictly forbade cosmic pluralism.

▲ A badass statue of Giordano Bruno near his execution site in Rome.

Bruno didn't care. He thought Copernicus was right. If anything, Copernicus hadn't gone far enough. Bruno was certain that we lived in an infinite universe filled with infinite stars orbited by infinite worlds inhabited with infinite alien life-forms.

"All those worlds contain animals and inhabitants no less than can our own Earth, since those worlds have no less virtue nor a nature different from that of our Earth," he wrote in his 1584 work, *On the Infinite, the Universe, and the Worlds*.

He did not view these positions as a threat to God's hegemony, but rather as a testament to His great and varied works. Some of Bruno's contemporaries might have also considered these possibilities; the difference was that Bruno could not help but talk about them, both openly and covertly, in addition to mouthing off about a range of other spicy topics considered heretical at the time.

As a consequence of his combative nature and unorthodox beliefs, Bruno spent his adult life bouncing around Europe, attracting and repelling patrons wherever he went. Eventually, he made the deadly mistake of returning to Italy, where he was promptly arrested by authorities and sentenced to execution for his many apparent heresies. Naturally, he refused to recant. In fact, he was such a boss that he told his prosecutors they were probably more afraid to dole out the sentence than he was to receive it.

He was burned at the stake in Rome's Campo de' Fiori on February 17, 1600. His works were banned and largely forgotten until, centuries later, he was recognized as a martyrial defender of free thought, a shift that revived his legacy.

Today, people from all around the world flock to Campo de' Fiori on the anniversary of Bruno's execution to pay their respects to the "Nolan philosopher." Bruno might have alienated a lot of these acolytes, too, if he were still around to get on their nerves. But while he has long since surrendered his corporeal form to the flames, Bruno's dream of a sacred, infinite, and living universe has survived him to this day.

> **"Thus is the excellence of God magnified and the greatness of his kingdom made manifest; He is glorified not in one, but in countless suns; not in a single Earth, a single world, but in a thousand thousand, I say in an infinity of worlds."**
>
> **—GIORDANO BRUNO**

# The Alien Content Boom

**T**HERE HAVE BEEN HINTS SINCE ANTIQUITY that people will go absolutely bonkers for stories about extraterrestrials (see Lucian's *True History* on page 10 and "The Tale of the Bamboo Cutter" on page 14). But nothing could prepare planet Earth for the explosion of interest in alien life that emerged in the Galilean age and, much like the ever-expanding universe itself, that obsession has only grown ever since.

In the decades following Bruno's death, infatuation with aliens continued to spread across Europe like a slow-rolling ground fire, producing stories about weird inhabitants of the Moon and other planets. Then, in 1686, a young man named Bernard Le Bovier de Fontenelle threw a bunch of gasoline on these flames with the publication of the first alien blockbuster.

### THE PICKUP ARTIST'S GUIDE TO ASTROBIOLOGY

With curiosity and charm to spare, Fontenelle had become a darling of the French intelligentsia by the time he published his masterpiece, *Conversations on the Plurality of Worlds*, at the tender age of twenty-nine. The book was inspired by his conviction that "nothing

Chapter One: ANCIENT ALIENS ▶ 021

could be of greater interest to us than to know how this world we inhabit was made, if there are other worlds which are similar to it, and like it are inhabited too," as he wrote in the preface.

To add a little spice to this topic, Fontenelle framed the book as a conversation between an inquisitive Marquise and a flirtatious male narrator who entrances her with cutting-edge astronomical research—*ooh la la*—as they stargaze over the course of five nights. (Then, as now, the best way to a woman's heart is through a lengthy review of contemporary theoretical cosmology.)

During the first few nights, the pair speculate about the nature of life on the Moon and in the solar system, resulting in memorable observations such as "Mercury is the lunatic asylum of the universe" and "Mars isn't worth the trouble of stopping there." On the fifth evening, the narrator and Marquise look beyond our solar neighborhood to the stars, which they conclude also host inhabited planets.

"Nature has held back nothing to produce [the universe]; she's made a profusion of riches altogether worthy of her," the narrator says. "Nothing is so beautiful to visualize as this prodigious number of vortices, each with a Sun at its center making planets rotate around it."

▲ Flirting under the stars, eighteenth century style.

Fontenelle not only widens the aperture of possible life to include the observable universe; he also anticipates the difficulty of detecting planets in other star systems, given that the immense glare of their host stars would drown out their own reflected light. This was just one of many important insights that Fontenelle popularized in his book, which also ruminates on Earth's appearance to alien life-forms and the question of whether aliens might look similar to or radically different from humans.

The persecution of Copernican pioneers, like Bruno and Galileo, hung heavy over the seventeenth century, and Fontenelle clearly feared he might also run afoul of religious authorities. To that end, he tried to get ahead of any blowback by directly addressing his clerical critics.

"When I say to you that the Moon is inhabited, you picture to yourself men made like us, and then if you're a bit of a theologian, you're instantly full of qualms," Fontenelle writes in the preface. "The descendants of Adam have not spread to the Moon, nor sent colonies there."

By emphasizing his view that aliens are not descended from Adam, Fontenelle hoped he might have found what he called a "loophole" that allowed open discussion of the topic without inviting disproportionate condemnation.

The gambit seemed to have worked. *Conversations* was a major success that reinvigorated support for cosmic pluralism across the Western world, and it did not provoke much objection from religious authorities. The Copernican cat was, at last, out of the bag. Fontenelle also encouraged his readers to envision worlds beyond Earth as physical entities with unexplored landscapes, valuable resources, and exotic lifeforms. Not bad for a stargazing cad.

# Aliens in the Scientific Revolution

**A**S THE SEVENTEENTH CENTURY WORE ON, scientists started to connect the dots between different subjects, as scientists are wont to do. Astronomical observations improved, revealing the cratered surfaces of the Moon and other planets—an anathema to Platonic predictions of perfect, featureless worlds.

Meanwhile, the dawn of microscopy exposed the existence of tiny living creatures, invisible to the naked eye, thriving all around us. These revelations suggested that there were countless niches where life could evolve beyond Earth, as well as an infinity of forms it might take.

▲ A 1697 woodcut of strange fireballs high above Germany. Humans have a long history of pointing at weird things in the sky.

# The Great Moon Hoax

**BREAKING NEWS:** The Moon is home to horned bears, biped beavers, and flying "man-bats" that have built amphitheaters and temples on the lunar surface. This thriving alien civilization was spotted with a novel telescope of vast dimensions that has since gone up in flames.

So goes the gist of a sensational 1835 report published in the *Sun*, a New York–based newspaper that credited the discovery of an extraterrestrial society on the Moon to Sir John Herschel, then one of the most famous astronomers in the world.

Known today as "The Great Moon Hoax," the series was the equivalent of clickbait for a nineteenth-century "penny press" paper. It was the brainchild of Richard Adams Locke, a distant relative of the philosopher John Locke, who set out to satirize astronomical discussions about alien life—and juice sales of the *Sun* in the process. But the prank went completely awry when readers missed the joke and took the "news" of this lunar civilization at face value.

The story ignited a media shitstorm that sent the *Sun*'s circulation into the stratosphere. Even Herschel initially got a kick out of it, though he eventually became annoyed at having to constantly repudiate the false account. The series also attracted criticism from Edgar Allan Poe, who accused the *Sun* of stealing his own attempt at a Moon hoax with a story called "The Unparalleled Adventure of One Hans Pfaall," which was published just two months earlier.

The Great Moon Hoax remains one of the most iconic examples of fake news in tabloid history, a decorated mantle to be sure. But it also exposed a credulous public hunger for tales of extraterrestrial life, and that appetite for aliens has only become more insatiable since Locke's self-described "aborted satire."

▲ The Moon's "man-bats" were described as "doubtless innocent and happy creatures, notwithstanding that some of their amusements would but ill comport with our terrestrial notions of decorum."

For scientists like Christiaan Huygens, a Dutch polymath born in 1629, all of this evidence amounted to the obvious existence of aliens, which he called "Planetarians" in his influential 1698 treatise *Cosmotheoros*. "Now should we allow the Planets nothing but vast Deserts," Huygens argued, "lifeless and inanimate Stocks and Stones, and deprive them of all those Creatures that more plainly speak their Divine Architect, we should sink them below the Earth in Beauty and Dignity; a thing that no Reason will permit."

Put another way, it's silly to conclude that all of these other planets, both in our solar system and beyond it, were inhospitable to life. What would be the point of them, then? As the seventeenth century gave way to the eighteenth, many other thinkers developed their own twists on this view, including Benjamin Franklin, Edmond Halley, William Herschel, Gottfried Leibniz, Immanuel Kant, and Thomas Paine.

As support for extraterrestrials flourished, people began to wonder how humans might stack up against these otherworldly species. I suspect Huygens would have been a Trekkie, as he dreamed of establishing interplanetary partnerships with other species in order to pool knowledge.

Others had a more self-effacing view of alien communications. David Rittenhouse, an American astronomer born in 1732, took a look around the world of his time and wondered why any intelligent species would want to establish communications with Earthlings—a common refrain to this day.

"We will hope that their statesmen are patriots, and that their kings (if that order of beings has found admittance there) have the feelings of humanity," Rittenhouse said, according to the essays of Founding Father Benjamin Rush. "Happy people!—and perhaps more happy still, that all communication with us is denied. We have neither corrupted you with our vices, nor injured you by violence. None of your sons and daughters have been degraded from their native dignity, and doomed to endless slavery in America, merely because their bodies may be disposed to reflect, or absorb the rays of light, different from ours."

As we see from this rather gloomy passage, opinions about aliens were beginning to more obviously reflect biases about life here on Earth. If you had a grim view of humanity, you might be a Rittenhouse-style pessimist about the prospect of alien contact. But if you yearned for a kinship between intelligent beings from all types of worlds, you might be more persuaded by Huygens-style alien optimism. The choice is yours!

▲ A satirical print depicting supposed Moon inhabitants after a lunar eclipse in 1724. Looks about right.

## William Whewell: Leader of the Alien Pessimist Minority

BY THE NINETEENTH century, a belief in alien life had taken hold across the Western world. But not everyone was convinced. One important skeptic was William Whewell, an English polymath born in 1794, who had shared hopes about extraterrestrial life in his younger days, but eventually came to doubt their existence for a variety of reasons.

In 1853, he anonymously published *Of the Plurality of Worlds*, a prescient takedown of alien optimism. In it, he posited that the rosy view of his peers was propped up by their existing spiritual and philosophical biases rather than on any hard evidence. Whewell, himself an Anglican priest, also had theological motivations for his work, but he scaffolded his argument with contemporary scientific findings. He pointed out that complex life shows up late in Earth's geological record, suggesting that even our own planet was relatively uninhabited for much of its history. He also anticipated the concept of a narrow habitable region around a star and noted the extreme conditions that likely existed on worlds like the Moon or Jupiter.

"We must either suppose that [Jupiter] has no inhabitants; or that they are aqueous, gelatinous creatures; too sluggish, almost, to be deemed alive, floating on their ice-cold water, shrouded forever by their humid skies," Whewell said.

Even today, the essential aim of astrobiologists and exoplanet hunters is to provide what Whewell called the missing "physical reasons" behind the assumption that aliens exist. The view that aliens exist is far more popular today (according to polls) than the idea that we are alone; still, it's important to note that nobody has actually proven Whewell wrong about his basic hunch that life may be uncommon in the universe.

▲ Sorry, not sorry: Whewell threw cold water on the alien optimism of the nineteenth century.

||||||||||||||||||||||||||||||||||||||||||||||||||||||||||||||||

We are intimately connected with
these faraway times and faraway places,
because it takes a whole cosmos
to make a human.

—JILL TARTER,
*astronomer and SETI pioneer*

||||||||||||||||||||||||||||||||||||||||||||||||||||||||||||||||

# CHAPTER TWO

# WHERE IS EVERYBODY?

# In the balmy late summer of 1924, people across the world prepared to receive the first messages from the intelligent alien civilization that was assumed to inhabit Mars.

**TO TUNE IN TO THESE MARTIAN COMMUNICATIONS, THE** US military imposed a period of radio silence on the nation that spanned 36 hours between August 21 and 23, coinciding with an unusually close pass between Mars and Earth in their orbits around the Sun. Martians, so the reasoning went, might capitalize on this moment of relative proximity—when the planets were a mere 34 million miles apart, instead of many times that distance—to reach out to their neighbors.

"Now, if ever, we may solve the disputed question whether superman rove [Mars's] crust," wrote the journalist Silas Bent in the *New York Times* on August 17, 1924. Hopes were high for contact with a Martian civilization, sparking "stupendous interest" from a "credulous public" in the lead-up to the historic Mars opposition, wrote the *Miami News*.

> **OPPOSITION:** The moment when two or more planets are positioned in a straight line in relation to their star. During a Mars opposition, Mars is on the exact opposite side of Earth relative to the Sun. Planets in opposition are roughly as close to Earth as they will be during any part of their orbit.

In addition to declaring a National Day of Radio Silence, the US military recruited Charles Francis Jenkins and David Peck Todd, a leading inventor and an astronomer respectively, to capture Martian missives sent across the interplanetary transom. For this purpose, Jenkins developed what he called a "radio photo message continuous transmission machine" that converted radio signals into optical flashes that could be recorded onto a 30-foot-long roll of photographic paper.

▲ Jenkins and Todd with the janky device they used to try to capture Martian messages.

When the opposition arrived at last, Americans flooded public observatories, hoping to catch a glimpse of their Martian neighbors. Jenkins and Todd, meanwhile, looked for messages recorded by the contraption in Jenkins's lab. The device did end up recording

an inscrutable arrangement of dots and dashes, a result that tantalized the public, though Jenkins did not believe he had made contact with Martians.

"Quite likely the sounds recorded are the result of heterodyning or interference of radio signals," Jenkins told the *New York Times*. "The film shows a repetition, at intervals of about a half hour, of what appears to be a man's face. It's a freak which we can't explain."

Other instruments turned up similarly puzzling results. An experiment in Dulwich, England, recorded "strange noises" early on the morning of August 23, which "could not be identified as coming from any earthly station," according to an Associated Press report published later the same day.

In the end, the tests turned up no concrete proof of life on Mars, let alone a civilization more advanced than our own. But that didn't sway public opinion much. People had been primed to believe in Martians thanks to a slew of purported evidence that had been championed by respected scholars in the decades leading up to the opposition.

Nikola Tesla, the polymath engineer, claimed to have intercepted Martian messages as early as 1899. In a splashy article published in a 1920 issue of *The Tomahawk*, Guglielmo Marconi, the Nobel

▲ As telescopes rapidly matured in the early twentieth century, astronomers were able to spot features on other planets and get a sense of the universe's massive scale.

Prize-winning engineer credited with inventing radio, also expressed hope that he had already made contact with aliens on Mars. And of course, the astronomer Percival Lowell was famously certain that he had spotted huge irrigation structures etched into the surface of the red planet, signaling the presence of an advanced, though possibly senescent, civilization. These observations were later determined to be optical illusions and perhaps some wish fulfillment on Lowell's part.

Press coverage of the topic regularly wove in skeptical snippets from other experts, but far more ink was spilled on the anatomies and abilities of the Martians that probably, likely, all-but-certainly existed. During this wobbly moment in history, haunted by the traumas of unthinkable war and unrest, people looked to Mars, hoping for a more mature alien society, like children seeking elder guidance.

"If there are beings on Mars, in the similitude of human beings, they are an order of intelligence much superior to ours," Silas Bent, the *New York Times* journalist, wrote in his preview of the Mars opposition. "Accepting that difficult premise, it is reasonable to suppose that the Martian knows much more about us than we know about him or his world, and it is interesting to speculate what he thinks of us, of our feverish struggle for a living, our vanities, our suicidal World War, our little gardens and our big deserts."

▲ Percival Lowell's (inaccurate) 1909 map of Mars depicted a complex system of canals.

# The Big Question

**T**HE BLIND FAITH IN ALIENS DURING THIS jubilant moment in August 1924 would get mercilessly body-checked by astronomical advances that unveiled an eerily silent universe. Even today, one hundred years later and armed with incredible technologies, we are reliving new iterations of this basic epiphany that life is simply not remotely as common—or at least, as easily detectable—as we once assumed.

▲ Ancient peoples speculated about life on the Moon, but missions to this world revealed an airless environment covered in hazardous dust.

The "seas" of the Moon turned out to be dry and sterile, and the "canals" of Mars were a mirage. The cloudy skies of Venus masked a hellscape below, not a lush paradise. At the same time, new telescopes revealed that the deep expanse beyond our solar system contained no obvious signs of other intelligences yearning for connection.

It was a harsh reality check. But not a dead end. If anything, the apparent absence of aliens in space supercharged their hold over the human imagination. As the dream of a bustling universe evaporated, humans simply insisted on seeing aliens in other places. Science fiction about aliens skyrocketed by the middle of the twentieth century, and hasn't wavered since. At the same time, a thriving worldwide subculture has emerged based around the conviction that aliens have already visited Earth and that shadowy forces have hidden their activities.

In the midst of this dizzying proliferation of fictional and unverifiable aliens, a small but dedicated band of scientists set out to search for real extraterrestrials guided by grounded investigative principles. We'll get to alien science fiction and ufology in Chapters Three and Four, but first we'll give a nod and perhaps some raised eyebrows to those early thinkers who sought to define the practical terms, approaches, and yardsticks in the search for alien life—and whose ideas still permeate the effort to this day.

## THE FERMI PARADOX

The feeling of inevitability about the discovery of alien life in the early twentieth century started to subside by the 1930s and 1940s, as astronomical technologies exposed inhospitable frontiers beyond Earth. The lack of clear evidence for alien life didn't necessarily imply that "they" weren't out there somewhere, but it did demand an explanation: Where is everybody?

Enrico Fermi, an Italian physicist and Nobel Prize laureate, accidentally became the historical avatar of this question when he blurted it out to a group of his colleagues over lunch. It was the summer of 1950, and Fermi was visiting Los Alamos Research Laboratory in New Mexico, where, just a few years earlier, he was a key member of the Manhattan Project, which developed, and detonated, the world's first nuclear weapon.

On that summer day, however, Fermi was idly chatting about the rising public reports of strange sights in the sky, dubbed unidentified flying objects (UFOs), with his colleagues Emil Konopinski, Edward Teller, and Herbert York. The subject was dropped for some time before Fermi, apropos of nothing, posed a version of the iconic question—where *is* everybody?—in the middle of their midday meal.

And with that random non sequitur, the legend of the *Fermi paradox* was born. In its simplest form, the paradox expresses the tension between the expectation that intelligent spacefaring aliens ought to be common in the universe and the observational reality that we haven't found, or been found by, any of them. Fermi died just a few years after that fateful lunch, but his paradox continues to puzzle people today and remains a fundamental concept in the search for life beyond Earth.

## THE TWIN PAPERS OF 1959

The Space Age would soon dawn as the *Sputnik* satellite launched in 1957, paving the first road off of Earth. Further fomenting excitement was the construction of sophisticated new telescopes, such as the one at the Green Bank Observatory in West Virginia, that dramatically extended our instrumental eyesight into the deep reaches of the Milky Way and beyond. A bit later in the decade, there must have been something special in the air in the autumn of 1959, when two unrelated scientific studies appeared in reputable journals. These papers would, in tandem, signal the dawn of the modern search for alien life.

"Searching for Interstellar Communications," a short work by physicists Giuseppe Cocconi and Philip Morrison at Cornell University, and "The Problem of Life in the Universe and the Mode of Star Formation," a study by astrophysicist Su-Shu Huang at the University of California at Berkeley, made considerable waves. Together, the studies began the process of translating Fermi's nagging question—where is everybody?—into a scientifically grounded research topic that could be constrained by, and perhaps even resolved with, contemporary technologies.

In their study, Cocconi and Morrison hypothesized that intelligent extraterrestrials might have already identified our solar system as

a potential site for life and that our planet could be awash in alien messages. But how might humans tune in to this interstellar chatter?

The researchers predicted that a message would try to incorporate some kind of universal mathematical truth, such as a string of prime numbers, to highlight its artificial origin. Cocconi and Morrison also introduced the idea that radio messages from aliens might be transmitted at what's known as the hydrogen line, also known as a 21-centimeter signal.

> **WHAT IS THE HYDROGEN LINE,** and why do we think aliens might chat on it? Hydrogen is the most abundant element in the universe, and a lot of it is just hanging out in space in a neutral state, meaning it has no electrical charge. Every once in a while, an electron in one of these neutral hydrogen particles changes the direction of its spin, a transition it marks by shooting out a little blast of light. These tiny flashes reach Earth as radio wavelengths with a distinct width of 21 centimeters, which corresponds to a frequency of 1,420 MHz. With this in mind, scientists have proposed that the sheer abundance of neutron hydrogen and the clear signature of these electron spin flips make this frequency an obvious channel for communication.

▲ Green Bank Observatory in West Virginia, the world's largest fully steerable radio telescope. Imagine taking this baby for a spin.

"Few will deny the profound importance, practical and philosophical, which the detection of interstellar communications would have," Cocconi and Morrison concluded in the study. "We therefore feel that a discriminating search for signals deserves a considerable effort. The probability of success is difficult to estimate; but if we never search, the chance of success is zero."

Whereas Cocconi and Morrison focused on searching for extraterrestrial messages, Huang wondered how to spot the home worlds of these speculative aliens in other star systems. In this way, he pioneered the field of exoplanet science long before it actually became possible to detect worlds beyond our Sun.

Indeed, you can almost sense Huang's frustration at being stuck in a time when exoplanets were a theoretical dream rather than an observational reality. Despite the technological limitations in 1959, Huang maintained an optimistic note about aliens by deducing that, based on available evidence, planets—and therefore, life—are likely common in the universe. As part of this effort, he imagined what kind of planetary and stellar environments might be most conducive to the emergence of life, and he coined the term *habitable zone* to describe the region around a star that is most likely to support life. He even predicted that a few nearby stars, Eridani and Tau Ceti, were likely contenders to host habitable planets.

> **BIOSIGNATURE:** A sign of biological activity, such as fossils, atmospheric gasses with biogenic origins, or chemical compounds linked to life.
>
> **TECHNOSIGNATURE:** A sign of technological activity, such as atmospheric pollution, interstellar messages, megastructures, or artifacts.

▶ Concept art of a planet and moon orbiting Epsilon Eridani, a young star that has been a hotspot of alien research and science fiction lore for decades.

## PRINCESS OZMA AND THE TWO STARS

In the same year that Cocconi, Morrison, and Huang began laying the theoretical groundwork of the search for extraterrestrial life, Frank Drake, an astronomer at Cornell University, was gearing up to conduct the first experiment to search for aliens with tools sophisticated enough to potentially find them: Project Ozma.

> **PROJECT OZMA** was named after Princess Ozma, the benevolent and rightful ruler of the land of Oz in L. Frank Baum's famous book series.

Drake was just shy of thirty, but had already racked up an impressive list of credentials as a radio astronomer, foreshadowing what would ultimately become his consequential role in the search for alien life. With the help of the (then) newly minted Howard E. Tatel Radio Telescope, an 85-foot dish at the National Radio Astronomy Observatory (NRAO) in Green Bank, West Virginia, Drake hoped to pick up radio signals at the hydrogen line from an intelligent civilization.

From April to July 1960, he aimed the dish at the Epsilon Eridani and Tau Ceti systems, the same sunlike stars singled out by Huang as most likely to host inhabited planets. In hindsight, these two stars were great targets for the study: We know now that Epsilon Eridani, located 10 light years from Earth, hosts at least one planet, while Tau Ceti, located 12 light years from Earth, has at least four planets, two of which are in the habitable zone of the system.

Ultimately, Project Ozma did not detect any alien messages, but the experiment generated a lot of scientific and public interest. It also directly led to a covert meeting of researchers, nicknamed the "Order of the Dolphin," at Green Bank in 1961 that has become legendary in the annals of the search for extraterrestrial intelligence (SETI).

## The Order of the Dolphin

**CONVENED BY J.P.T. PEARMAN** of the National Academy of Sciences, the Green Bank meeting aimed to figure out how to communicate with alien life-forms. We've already met some of the attendees—Drake, Huang, and Morrison—but the meeting also included radio enthusiast Dana Atchley, biochemist Melvin Calvin, computer scientist Barney Oliver, counterculture researcher John Lilly, and the young astronomer Carl Sagan (before he had made his name as a beloved turleneck-wearing science communicator).

The group met over the week of Halloween to define the major challenges of the nascent SETI field and debate the best approaches to meet them—all while finding some time to pop some breakfast bubbly for Calvin after he won the 1961 Nobel Prize in Chemistry.

In addition to sketching out the rudiments of SETI research, the men evidently had a lot of fun together. After repeatedly discussing the difficulties of opening communications with dolphins—a clearly intelligent and chatty species right here on Earth—the members of the group playfully named themselves the "Order of the Dolphin." Calvin even made dolphin pins to memorialize the conference.

### THE DRAKE EQUATION

With no disrespect to their novelty cetacean ornaments, the most consequential product of the Dolphins' 1961 meeting was no doubt the Drake equation, a string of variables that has since been endlessly invoked and debated in SETI circles. Frank Drake presented his equation to the Dolphins as a means to pull the fledgling field out of nebulous ambiguity and pin down the odds of alien contact with some real-world variables.

So how does it work? The equation is designed to estimate the number of extraterrestrial civilizations with which humans might feasibly communicate (represented by $N$). Drake tried to home in on the major factors that might influence the possible detection of alien life, such as the average number of potentially habitable worlds or the lifespan of advanced civilizations:

$$N = R^* \times f_p \times n_e \times f_l \times f_i \times f_c \times L$$

where:

- $N$ is the number of civilizations in our galaxy capable of communicating with Earth
- $R^*$ is the annual rate of formation of stars that could conceivably host life
- $f_p$ is the fraction of those stars that have planetary systems
- $n_e$ is the average number of planets per star capable of supporting life
- $f_l$ is the fraction of these planets that actually develop life
- $f_i$ is the fraction of inhabited planets that develop intelligent life
- $f_c$ is the fraction of civilizations that develop a technology that releases detectable signals
- $L$ is the length of time these civilizations communicate

Back in 1961, none of the variables in the Drake equation were well constrained by empirical evidence. We are lucky, more than sixty years later, that some gaps have been filled: Scientists currently estimate that about ten stars are born in our galaxy each year, and that there is an average of at least one planet for every star system, putting the total number of planets at, conservatively, 400 billion in the Milky Way. The rest of the variables remain speculative, resulting in estimates that predict that there are hundreds of thousands of technological civilizations in our galaxy on the high end, all the way down to an emphatic zero on the low end.

In other words, the Drake equation is highly malleable and reflects the biases of its practitioners. Some scientists have critiqued the formula's shaky conjectural foundations and produced their own modifications. With that in mind, it's important to note that Drake never intended for his equation to produce accurate estimates. Rather, it was an attempt to sketch out a rough road map that might come into sharper focus in tandem with new scientific discoveries.

## KARDASHEV ENTERS THE CHAT

The Dolphins weren't the only ones mulling over the existence, and potential detection, of alien life during the 1960s. Nikolai Kardashev, born in Moscow in 1932, was a SETI pioneer who was about the same age as Drake and Sagan when he began to lead similar efforts to advance the search for extraterrestrials in the Soviet Union.

Kardashev organized his own alien-curious conference at the Byurakan Astrophysical Observatory in Armenia in 1964. The Soviet researchers who gathered there were all aware of Project Ozma and the general conclusions of the Dolphins. But in contrast to Drake's tight focus on two stars with Ozma, Kardashev believed astronomers should search for incredibly high-powered signals from unthinkably advanced civilizations.

At the 1964 conference, Kardashev unveiled a famous ranking system that now bears his name. The Kardashev scale groups hypothetical alien civilizations into three categories that are divided by energy consumption:

- A Type 1 civilization can harness all the energy available on their home planet.

- A Type 2 civilization can access all the energy of their host star.

- A Type 3 civilization can get drunk on the power of their entire galaxy.

Since our own human society falls short of being considered a Type 1 civilization, Kardashev thought we would have better luck looking for Type 2 or 3 civilizations that could tap into the energies of stars and galaxies.

"If terrestrial civilization is not a unique phenomenon in the entire universe, then the possibility of establishing contacts with other civilizations by means of present day radio physics capabilities is entirely realistic," Kardashev noted in a 1964 study that summarized his scale. "At the same time, it is very difficult to accept the notion that of all of

## Project Cyclops

**IN 1971, NASA** launched Project Cyclops, a lengthy study investigating how best to conduct SETI research going forward. Over the course of a 250-page report, the project team outlined an ambitious plan to build thousands of 100-meter-wide dishes that could detect technological signals across a radius of 1,000 light years from Earth.

The report acknowledged that this effort would cost several billion dollars, a financial reality that doomed the Cyclops vision from ever materializing. In 1971, NASA's budget was heavily strained by the Apollo program, which was still actively landing astronauts on the freakin' Moon. There was no change left in the proverbial couch cushions, especially for a project that many people considered to be a wild extraterrestrial goose chase.

Although Project Cyclops never came to fruition, it established influential guidelines for future iterations of SETI research while also capturing a fascinating snapshot in time. The authors of the report justified the search for alien civilizations with detailed scientific findings, though they conceded that any human attempt to make contact with intelligent beings was an act of faith. "Faith that the quest is worth the effort," the team said in the report, "faith that man will survive to reap the benefits of success, and faith that other races are, and have been, equally curious and determined to expand their horizons."

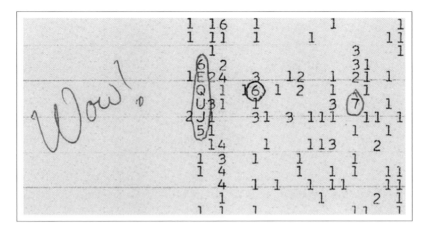

the [billions of] stars present in our galaxy, only near the Sun has a civilization developed. It is still more difficult to extend this inference to the [billions of] galaxies existing in the portion of the universe accessible to observation. In any case, the deciding word on this question is left to experimental verification."

## THE WOW! SIGNAL

Consider the following sequence of characters: 6EQUJ5. At first glance, it looks like something innocuous—perhaps a license plate or a postal code—but it is, in fact, a representation of a bizarre radio signal from outer space that remains the most compelling candidate for an alien message in history.

The jumble of numbers and letters expresses the intensity pattern of a narrowband radio signal captured by the Big Ear telescope in Delaware, Ohio, on August 15, 1977. The event lasted 72 seconds and seemed to originate from the direction of the constellation Sagittarius. It was also a distinctly loud signal that reverberated close to the hydrogen line, matching the SETI community's predictions about likely communication frequencies.

▲ A printout of the Wow! signal. The key measurement is the "U" which expresses that this signal was 30 times more intense than background radiation.

It took a few days for anyone to notice the tantalizing blast in the dataset. But when astronomer Jerry R. Ehman first saw the profile of the signal, he was so taken aback that he circled the "6EQUJ5" sequence on a printout and wrote "Wow!" in the margin. Thanks to this enthusiastic scribble, the event is now known as the "Wow! signal."

The origin of the Wow! signal still remains a mystery nearly fifty years after Big Ear captured it. Though it had some of the expected features of an alien message, it was never picked up again, suggesting that it was a one-off event. If an alien civilization were interested in making contact, it stands to reason that they might repeat their signals to better distinguish them as artificial.

Then again, humans have sent a smattering of weird messages into space in all directions that don't add up to any kind of coherent whole and may well be prompting alien civilizations to say their version of "What the heck?" at this very moment. (See a sampler of bizarre greetings for our alien friends on page 48.)

> **ACTIVE SETI:** Any attempt to send messages to extraterrestrial civilizations, also known as Messaging to Extraterrestrial Intelligence (METI).

## *SHOULD* WE TRY TO CONTACT ALIENS?

In January 1929, the British *Manchester Guardian* newspaper asked its readers to respond to the following question: "What message would you send to Mars if an interplanetary wireless service were put into operation?"

Some responses were highly practical; for instance, a seven-year-old reader hoped to ask, "Do you eat?" Other responses reflected controversies of the times: A reader asked about the job market on Mars, noting that unemployed Britons could be useful canal-diggers.

Another wondered if evolution was taught in Martian schools (the Scopes Monkey Trial had captivated international attention just a few years earlier).

For the most part, readers seemed hopeful that Martians might be able to help Earthlings build a better world. This optimism that aliens could cure our human ills and benefit our civilization continued to find a voice in the following decades. In an article published in 1964, author Isaac Asimov even declared his belief that establishing "interstellar communication would be to bring about a contact of minds that can result only in good and not evil."

"Why ought we to be so certain that an intelligent alien would find nothing better to do than to destroy us?" Asimov asked in the piece.

Of course, the obvious rejoinder is that we can't be certain of anything intelligent aliens would do, so why risk calling attention to ourselves? Is it really a good idea to shout out into the cosmic dark when we have no idea what or who is hiding in the shadows?

Spoiler alert: People have gone ahead and tried to contact extraterrestrials anyway, despite these concerns. But the debate over the wisdom of actively seeking contact with aliens, as opposed to passively detecting them from Earth, has only intensified over the past century.

▲ The Arecibo Message, broadcast into space in 1974, includes the numbers one through ten in white at the top, a stick figure of a human in red, and the Arecibo Telescope in purple at the bottom. It's the SETI version of the age/sex/location chatroom prompt.

# A Short History of Alien Icebreakers

**H**UMANITY SHOUTED ITS FIRST MESSAGE FOR aliens into space from a glorious network of dishes in Crimea that was partially spun together from spent weaponry—literally submarine hulls and battleship gun turrets—giving it a truly unique and unruly appearance. The date of the call was November 19, 1962. The target was the planet Venus. The language was Morse code. The message was "MIR, LENIN, USSR."

A response never came, of course, but the designers of this experiment hadn't really expected to make contact with intelligent Venusians who wanted to bone up on Bolshevik history. The effort, now known as the Morse Message, was intended as a test of what was then known as the Yevpatoria Planetary Radar complex. But it also kicked off a series of ever-evolving attempts to transmit a variety of different greetings into space, including...

∷ THE PIONEER PLAQUES: A pair of identical visual plaques on board NASA's *Pioneer 10* and *11* probes, launched in 1972 and 1973, that are currently speeding out of the solar system.

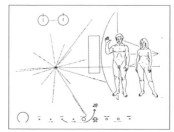

▲ The Pioneer plaques included representations of male and female anatomy, the hydrogen line, and the solar system's configuration and position in the Milky Way.

::: THE ARECIBO MESSAGE: A three-minute-long radio message, written in binary digits and transmitted from the Arecibo Observatory in Puerto Rico on November 16, 1974. The transmission was directed at a bundle of stars known as Messier 13, which it will reach in about 25,000 years. The Arecibo Message, like the Morse Message, was intended more as a demonstration of human technologies than as an earnest attempt to contact aliens.

::: THE VOYAGER GOLDEN RECORDS: A pair of identical records that carry images, music, and other earthly offerings on board NASA's *Voyager 1* and *Voyager 2* probes. The *Voyager* probes were launched in 1977, and both have now ferried their Golden Records into interstellar space.

::: "POETICA VAGINAL": Sounds of vaginal contractions from dancers at the Boston Ballet, transmitted by the artist Joe Davis using MIT's Millstone Hill Radar in 1985. Yes, you read that right—vaginas have gone interstellar. Though Davis's experiment was shut down by the United States Air Force within minutes, the truncated recordings reached the stars Epsilon Eridani and Tau Ceti in the 1990s, and they still continue to spread the good word of earthly genitalia to the wider galaxy.

::: TEEN AGE MESSAGE: A series of interstellar radio messages, sent from the same complex in Yevpatoria, in 2001. The target stars and contents of the message were decided in collaboration with a group of Russian teenagers, who made the excellent call to include a theremin concert in the broadcast.

▲ The cover of the Voyager Golden Record includes instructions for playing it. Aliens definitely dig vinyl.

::: **CRAIGSLIST IN SPACE:** Thousands of posts on Craigslist, the community advertising website, were transmitted into space on March 11, 2005, by a company called Deep Space Communications Network. So if an alien shows up to buy your ratty futon from college, blame this message.

::: **ACROSS THE UNIVERSE:** An interstellar radio message containing the Beatles's song "Across the Universe" transmitted to the star Polaris by NASA from a 70-meter dish near Madrid, Spain, on February 4, 2008. It's a pretty mid-grade Beatles song to send into space, but the best one (in my opinion), "Happiness Is a Warm Gun," could conceivably be viewed as a threat.

::: **DORITOS AD IN SPACE:** A short advertisement for Doritos beamed as an MPEG file from the Arctic Circle in the direction of the constellation Ursa Major on June 12, 2008. It was the first advertisement ever directed to potential extraterrestrial life, so don't be surprised if we are invaded for our Cool Ranch resources.

::: **A MESSAGE FROM EARTH:** An interstellar digital radio signal containing hundreds of crowdsourced messages, directed at the Gliese 581c system, was sent from the Yevpatoria complex on October 9, 2008. The message is considered the world's first publicly curated digital time capsule and will reach its target in 2029.

::: **WOW! REPLY:** An interstellar message sent from Arecibo Observatory on August 15, 2012, was developed by National Geographic as a response to the Wow! signal of 1977. The reply included crowdsourced tweets and celebrity contributions, including a video of Stephen Colbert discouraging aliens from eating humans on account of our gamy meat.

**LONE SIGNAL:** A series of short messages, transmitted, toward the Gliese 526 system, beginning on June 17, 2013. Gliese 526 is about 17 light years from Earth, meaning that it takes 17 years for light to travel from here to there (one light year is equivalent to about 5.88 trillion miles). The project was funded by entrepreneurs who invited the public to send their own messages to space for 25 cents per transmission, but it shut down shortly after it launched due to a lack of funding and interest.

**A SIMPLE RESPONSE TO AN ELEMENTAL MESSAGE (ASREM):** A fourteen-minute interstellar radio message transmitted on October 10, 2016, was encoded with thousands of responses to the question: "How will our present environmental interactions shape the future?" People from all around the world contributed their thoughts, which were beamed out to the North Star, Polaris, by a European Space Agency facility near Madrid, Spain.

**LINCOS:** A portmanteau of *lingua cosmica*, Lincos is a constructed language developed by the mathematician Hans Freudenthal in the 1950s that is designed to communicate with alien life. He created it using mathematical principles under the assumption that an alien capable of receiving a radio transmission would probably be, first and foremost, a math nerd.

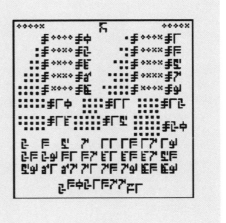

▲ Lincos: A language designed for extraterrestrial communication (and cross-stitch projects).

# The Search for the Origin of Life on Earth

**WHEN I THINK** about abiogenesis, the term that describes the emergence of life from nonliving materials, I'm often reminded of an iconic early episode of *South Park* called "Gnomes." The episode featured tiny enterprising gnomes who steal underpants as part of a three-part business plan: Phase 1) Collect underpants. Phase 2) ? Phase 3) Profit.

The quest to answer how life first emerged on Earth seems to follow a similar basic scheme: 1) Collect ingredients for life. 2) ? 3) Life.

This analogy is not meant to cast any shade on the wealth of research into the mind-boggling mystery of how a bunch of dirt and water managed to arrange itself, over time, into a host of self-sustaining, self-replicating living cells. In recent decades, scientists have made huge strides in untangling this problem by chasing elusive clues in the geological record, or peering at fundamental biological processes in laboratory experiments, or generating models that simulate possible primordial cellular and metabolic processes. Still, though, that second phase between not-life and life remains a giant question mark.

Here's what we do know: Life got started on Earth pretty quickly, within a billion years of our planet's birth. The environment at this time would have been deadly to humans and most other forms of modern life, as there was no significant atmospheric oxygen and a much more intense bombardment of asteroids and comets than we experience today. Indeed, ancient life may have been sparked many times in many places, only to be wiped out, over and over, by these initially hostile conditions.

The earliest living organisms may have emerged deep in the oceans; around volcanic seafloor structures called hydrothermal vents; in carbon-rich lakes on the surface; or in some place we have never even considered. Regardless of where the transition happened, photosynthetic life-forms entered the scene around 3.5 billion years ago and completely reshaped the planet. As these organisms produced more and more oxygen, a byproduct of photosynthetic reactions, they shifted the composition and dynamics of the atmosphere. Any life-forms that couldn't live under oxygenated skies were wiped out, but those that made it through were gifted

with a more stable climate. Eventually, these conditions led to the rise of complex life, including animals and plants, about 600 million years ago.

What does this sketch of Earthling history tell us about the odds of non-Earthling life? That is an evolving question with answers that span all kinds of fields. But one interesting takeaway is that life may create the conditions for its own survival, or its own doom, without even being aware of it. The over-under on that process could determine whether our long-lived biosphere on Earth is an aberration or the norm.

> "In the beginning, there was nothing, which exploded. A lot of this became stars. Some of it became rocks. Some of it became a sort of gooey liquid. Some of it sat around and got itself a name. This was called life. And then it had lunch."
>
> —TERRY PRATCHETT, *The Last Continent*

▲ Photosynthetic life in the oceans completely changed Earth's climate, paving the way for complex ecosystems.

||||||||||||||||||||||||||||||||||||||||||||||||||||||||||||

We have no need of other worlds.
We need mirrors. We don't know
what to do with other worlds.
A single world, our own, suffices us;
but we can't accept it for what it is.

—STANISŁAW LEM, *Solaris*

CHAPTER THREE

# POP CULTURE ALIENS

# Nothing is more human than imagining the alien.

**JUST TAKE A MOMENT TO STEP BACK AND CONSIDER THE** breathtaking abundance and variety of extraterrestrial beings that turn up everywhere from cheap pulpy comics to blockbuster Oscar-winning films. While aliens may not have literally invaded Earth—at least, not that I'm personally aware of—there is no doubt that their apparitions have thoroughly colonized our brains.

And why not? Aliens are narrative dynamite. They can take any form, drive any plot, advance any message. They can be lowbrow. They can be cerebral. They can be villains. They can be heroes. They can be totally inscrutable. They can be just like us.

Yet despite the dazzling diversity of alien tales, there is one clear throughline: All of our fictional aliens are made to illuminate humanity. In the absence of any real experience with extraterrestrial life, we have come up with a whole universe of possible windows through which to study ourselves.

This imaginative exercise dates back thousands of years, as we learned in Chapter One. But it has experienced a steep change in creativity and popularity over the past 150 years, propelled by both advances in science and the maturation of speculative fiction as a distinct genre. It's as if we are now in the midst of a Cambrian explosion of pop culture aliens, as new versions of this ancient archetype mutate, diversify, and spread across our cultural substrates.

These imagined extraterrestrials take many forms. They almost always fall into at least one of five thematic clades: Monsters. Saviors. Peers. Subordinates. Strangers. (And maybe a few others, too—exodinosaurs included.) Let's take a closer look at why these characters have so successfully proliferated through our mental landscapes.

# Monsters

> "We must remember what ruthless and utter destruction our own species has wrought. . . . Are we such apostles of mercy as to complain if the Martians warred in the same spirit?"
>
> —H. G. WELLS, *The War of the Worlds* (1898)

EXTRATERRESTRIAL MONSTERS, IN THEIR modern form, owe much of their heritage to H. G. Wells's watershed novel *The War of the Worlds*, which has been hitting nerves ever since it was first published in 1898.

On its face, the story is about murderous Martians that invade Earth for its resources while consuming their human and animal victims with voracious intensity. But Wells, an Englishman living at the cusp of the twentieth century, was unsettled by the excesses of his own nation's empire and its devastating effects on wildlife and people. So, while he clearly relishes describing the gruesome acts of the invaders, Wells also asks his readers to look in the mirror to an almost obsessive degree. The Martians' bloodthirst is "no doubt horribly repulsive to us," says Wells's narrator, "but at the same time I think that we should remember how repulsive our carnivorous habits would seem to an intelligent rabbit."

▲ The Martians in *The War of the Worlds* are gelatinous octopus-like creatures, but they rampage across Earth in killing machines called tripods.

Grossed out by the black-eyed, sharp-toothed, blood-guzzling aliens? That's what *you* look like to a cute little bunny, Wells insists. This type of self-reflective commentary pops up in all sorts of alien monster tales, especially in the subgenre of military science fiction. Orson Scott Card's 1985 novel *Ender's Game* introduces alien foes that initially seem monstrous—before the script gets flipped to reveal staggering deceit and cruelty from the human "protagonists."

Similarly, the 1997 film *Starship Troopers*, adapted from Robert Heinlein's 1959 novel of the same name, turns the murky themes of its source material up to bombastically ironic levels with its vision of a joyfully fascist state bent on destroying alien "bugs" with excessive war tactics and brainwashing campaigns pushing a blunt-force message: "Violence is the supreme authority."

> **THE MONSTROUS MOTHER:** *Ender's Game* and *Starship Troopers*, along with the Alien franchise, exemplify this excellent alien monster subcategory. Monstrous Mothers are depicted as grotesque breeders shot through with deadly maternal instinct, giving them a visceral and almost relatable dimension that is absent from more generic monsters.

There are also plenty of alien monsters in science fiction that are just supposed to be scary villains, not necessarily avatars for human shortcomings. While the 1996 blockbuster *Independence Day* owes some of its heritage to Wells's novel, it also offers an unapologetic celebration of American jingoism with no irony or self-reflection in sight to dampen the mood.

And there's need to psychoanalyze a character like the Blob, an alien monster that has starred in two eponymous films and engorges itself on whatever flesh it encounters (though you could probably come up with a metaphor about consumerism or something if you had to write a term paper). At the end of the day, sometimes it's just fun to have a big freaky beast to fight, and that beast occasionally hails from outer space.

# Saviors

> "Live as one of them, Kal-El, to discover where your strength and your power are needed. Always hold in your heart the pride of your special heritage. They can be a great people, Kal-El; they wish to be. They only lack the light to show the way. For this reason above all, their capacity for good, I have sent them you: my only son."
>
> —JOR-EL, *Superman Returns*

A VERY CLEAR MESSIANIC THROUGHLINE RUNS through humanity's obsession with aliens. Even many scientists can't help but speculate about the wisdom and knowledge that alien contact could bestow on humans. No character embodies this thread better than Superman, a monolithic presence in popular culture since he debuted in 1938 as a car-brandishing strongman in the iconic *Action Comics #1*.

Superman was created by Jerry Siegal and Joe Shuster, both of whom were first-generation North Americans born to Jewish refugees. It's no accident that Superman experienced the loss of his ancestral homeland, the planet Krypton, and had to adapt to the customs

of a new land, Earth. This origin story has profoundly resonated with audiences across the decades, as Superman struggles to live up to his destiny as a protector and aspirational figure for humankind, even though he is not, and never will be, One of Us.

The Man of Steel has a host of amazing abilities, such as flight, X-ray vision, super-strength, and super-speed, but his essential power is a transcendent goodness that has continued to appeal to new generations. Christopher Reeve, the late actor who played the character in several films, noted that "what makes Superman a hero is not that he has power, but that he has the wisdom and the maturity to use the power wisely." This observation is borne out by the fact that many of Superman's greatest foes are fellow Kryptonians with commensurate powers that they use to diabolical ends (looking at you, Zod).

▲ Superman is popular culture's most popular example of an alien savior.

Superman and other alien saviors are sometimes overtly depicted with Christian iconography, and a classic messianic message runs through many of these stories. The extraterrestrial Klaatu from the 1951 film *The Day the Earth Stood Still* is often interpreted as an alien Jesus on account of his narrative arc, which features martyrdom and resurrection, as well as his warning that human civilization will be destroyed if we fail to curtail our avarice and destructive impulses.

In an interesting twist on the theme, Thanos, a swole alien and the main villain of the Avengers film franchise, sees himself as a savior who must sacrifice half of all life in the galaxy, including his own loved ones, to ensure a brighter future for the survivors. Needless to say, his commitment to this vision is not shared by the film's heroes. Frank Herbert's Dune series offers a similar contortion of the traditional messianic arc, but while these tales take place on extraterrestrial planets, they are not explicitly alien science fiction, as the main characters are all part of the same human diaspora (with the exception of the big ol' sand worms and other native wildlife).

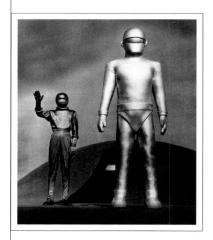

Alien saviors also aren't always saddled with the cumbersome job of redeeming *all* of humanity; sometimes, they liberate only a single person. Steven Spielberg's most iconic alien films, *Close Encounters of the Third Kind* (1977) and *E.T. The Extraterrestrial* (1982), are both centered on alien visitations that provide solace and purpose to individual people who are struggling with fairly pedestrian earthly circumstances. But whether they set out to serve all of humanity or just one individual, alien saviors have benevolent intentions and the power to bring out the best in our species.

▲ Klaatu emerges from his spacecraft with his backup robot, Gort, who can shoot lasers out of his face, in *The Day the Earth Stood Still*.

As a consequence, this alien trope carries all of the baggage of its parent family of messianic narratives that have appealed to mass audiences for thousands of years. On a basic level, these stories are aspirational salves for a species that seems to be constantly kicking its own ass. Given how prone humans are to enacting and perpetuating suffering on ourselves and other Earthlings, wouldn't it be nice to just have world peace delivered to us on an extraterrestrial platter?

The downside of this desire for an *alien ex machina* moment is that it portrays humans as unable to clean up our own messes without the intervention of an altitudinally higher power. In his 1987 essay "The Alien Messiah," the scholar Hugh Ruppersburg argues that the cinematic iterations of this trope thus reflect reactionary, defeatist attitudes.

"If they do not reject science and technology, they at least ignore it," Ruppersburg writes. "If they regard the future with hope and wonder, they simultaneously discourage the hope that humankind will be more capable in the future of handling the problems that confront it today. Entertaining as they are, these films are escapist fantasies grounded in the patterns of the past instead of the possibilities of the future."

Whether or not you agree with this assessment, alien saviors are likely to stick around for as long as humans desire divine deliverance from our terrestrial misadventures. And anyway, there's nothing wrong with a little escapism now and then.

▲ E.T. brings the savior trope to a personal level through his bond with human pal Elliott.

# Peers

> "I have been, and always will be, your friend."
>
> —SPOCK, *The Wrath of Khan*

ALIENS: THEY'RE JUST LIKE US. OR, AT LEAST, this is more or less the message of franchises like Star Trek that depict traditional alliances and rivalries between humans and other intelligent aliens. In these stories, human and alien characters are roughly equivalent in terms of their knowledge and technologies—and occasionally anatomically compatible (wink wink). The narrative tension is often driven by diplomatic efforts to bridge cultural divides and avert conflict in both interpersonal and civilizational contexts.

Extraterrestrial peers tend to present a more optimistic vision of human-alien contact—and sometimes human-human contact—than other categories, even when evil forces are ascendent in their universes. Part of the lasting appeal of the Star Wars franchise, for instance, is the ragtag makeup of the Rebel Alliance, which includes beloved alien characters like Chewbacca, Yoda, and Admiral Ackbar battling the Empire alongside human characters. These examples carry an implicit message that monomaniacal villains can be defeated by alliances that are forged from diversity and embody tolerance for other cultures.

While these stories about extraterrestrial peers are often modeled on the real challenges of international relations here on Earth (see "Galactic Geopolitics," page 75), they can also offer intimate portraits of interpersonal connection. To that end, aliens and humans become close pals, as with Elliott and E.T., Han Solo and Chewbacca, Kirk and Spock, and countless other iconic friendships depicted in fictional worlds heavily populated by aliens (the Marvel universe, too many manga books to count, and so on).

But just as often, humans and aliens end up as decidedly more than friends. As we saw in Lucian's *True History*, human-alien sex is basically as old and varied as alien fiction writ large, and it opens an immensity of possibilities to elevate human horniness to cosmic heights.

Often, interspecies dalliances are played for laughs, as with the 1988 comedies *Earth Girls Are Easy* and *My Stepmother Is an Alien* or the television series *Third Rock from the Sun*. But there is also a voluminous corpus of romances and erotica about aliens and humans who come up with all kinds of inventive and ingenious ways to use their diverse anatomical configurations to satisfy their interplanetary lust.

▲ Captain Kirk (William Shatner) and Commander Spock (Leonard Nimoy) are interstellar BFFs and have the banter to prove it.

> **SIDE NOTE:** Alien-human sex is not just limited to loving peer stories; predatory and nonconsensual sex also occurs across all kinds of alien tales.

There's sex via prehensile braided hair in *Avatar*. There's telepathic sex mediated by horny alien dragons in Anne McCaffrey's Dragonriders of Pern series. There's a full-on planetwide orgy when insatiable aliens endowed with a multitude of orifices and phalluses flock to Earth in Harlan Ellison's short story "How's the Night Life on Cissalda?" Video games have also allowed players to experience interspecies fornications; in the Mass Effect series, the playable human character can choose to hook up with any number of alien crew members (I'll never forget you, Garrus Vakarian).

Peer aliens are by far the most abundant type in science fiction. This proclivity to imagine aliens as our friends, lovers, rivals, and intellectual equals betrays our ancient longing to find companionship, of one form or another, somewhere in the stars.

▲ Jeff Goldblum as a blue hairy alien with formidable sexual prowess in *Earth Girls Are Easy*.

# Subordinates

> "When dealing with aliens, try to be polite, but firm. And always remember that a smile is cheaper than a bullet."
>
> —TRAINING VIDEO NARRATOR, *District 9*

HUMANS ARE QUITE PRACTICED AT DREAMING up advanced aliens that may pose threats and/or benefits as a consequence of their superior knowledge and power. However, there is a fascinating subcategory of stories that cast humans as the power brokers in their relationship with aliens.

Sometimes, these lopsided dynamics are presented as exploitative; in the 2009 film *District 9*, refugee aliens seek asylum on Earth but end up trapped as second-class citizens in brutally policed slums. Set in Johannesburg, South Africa, the movie clearly channels the nation's history of apartheid and xenophobia to achieve its heartbreaking narrative ends.

Subordinate aliens also turn up in lighthearted roles, often as pets or comic relief. These alien pets have a storied pedigree that dates back to Woola, a loyal Martian dog analog (anadog?), who appears in Edgar Rice Burroughs's Barsoom series published in the early twentieth century. The 2002 Disney movie *Lilo & Stitch* features a young girl who adopts a chaotic alien and tries to pass it off as her dog, leading to delightful lines such as "He used to be a collie before he got ran over," and "Oh good! My dog found the chainsaw!"

There is even a subcategory of this subcategory that features subordinate aliens that are secretly superior to humans. My favorite example of this little niche is an episode of the television series *Futurama* entitled "The Day the Earth Stood Stupid," in which the cycloptic character Leela discovers her cuddly companion, Nibbler—a winner of "Dumbest Pet in Show!"—turns out to be an ambassador from a super-intelligent species sent to monitor humans . . . in exchange for an abundance of hams and belly rubs.

▲ The aliens in *District 9* face mockery, derision, and cruelty from humans who see them as subordinate.

# Strangers

> "It's not like us. It's unlike us. I don't know what it wants, or if it wants."
>
> —DR. VENTRESS, *Annihilation*

**A**LIENS IN ALL OF THE PREVIOUS CATEGORIES we've explored display some basic human characteristics and motivations, such as a desire for conquest, connection, or protection. Alien strangers, however, are specifically designed to resist anthropomorphization. These aliens are, like, *alien* aliens.

*Solaris*, a 1961 novel by Stanisław Lem later adapted into an Andrei Tarkovsky film, is a pioneer of the stranger subgenre. The alien in this story is a vast, planetwide sentient ocean that is monitored by a human crew intent on establishing communications with the megabeing. The ocean responds to these efforts by producing ghostlike simulacra of the crews' loved ones for reasons that are never explained, leaving the human characters with more questions than answers.

There are echoes of *Solaris* in many subsequent stories about human scientists who are confronted with alien strangers. In the eerie 1982 John Carpenter film *The Thing*, a research crew in Antarctica

descends into paranoia after they are infiltrated by an alien that can imitate anyone. Jeff VanderMeer's Southern Reach Trilogy also explores an extraterrestrial entity that is able to produce new forms of life, including human dopplegängers.

Strangers often show up in stories centered on the limits of human understanding in an intractable universe. This notion that humans might reach the ceiling of our own comprehension has deep roots in many religious and philosophical movements. You can see it in the cross-cultural trope of "divine madness," or in the unknowable motivations of deities, as expressed in the popular eighteenth-century hymn "God Moves in a Mysterious Way."

These tales offer new vectors for more ancient themes about existentialism, transcendence, or the meaning of life (assuming there is one). They also defy popular expectations that aliens might provide insights, friendship, or any sense of recognizable connection; the stranger trope suggests that, sadly, contact with aliens may not provide the cosmic kinship or grand answers that we seek.

▲ The Thing, doing its thing.

# Other Alien Tropes

**I**N ADDITION TO THE BIG FIVE ARCHETYPES, alien science fiction has also adopted, and sometimes pioneered, countless tropes that have their own distinct evolutions and manifestations. Here is a modest sampling of popular narrative and thematic staples in stories about extraterrestrials, from galactic geopolitics to time distortion.

## HUMAN-ALIEN SHAPESHIFTERS

Many fictional aliens are depicted as anatomically humanoid, even if they have a few exotic flourishes to distinguish them as not of this world. This resemblance to humans makes alien characters more empathetic, and it is also a practical matter for film or television creators who don't have the budget for exotic costumes and effects. *Star Trek* famously demonstrated that you can slap a pair of pointy ears or some forehead ripples onto an actor and the audience will generally do the rest of the imaginary legwork.

However, there is a particularly rich subcategory of aliens that are *not* naturally humanoid, but have the ability to change into a human form

Reptilian humanoids are a mainstay of alien science fiction. ▲

for a specific purpose. Many of these shapeshifters follow a rich tradition of stories about supernatural entities, such as gods and demons, that walk among us on Earth, disguised and undetected.

This particular trope has been around for centuries, but it exploded into the mainstream in the wake of the 1956 horror film *Invasion of the Body Snatchers*. The plot follows an extraterrestrial invasion of "pod people" who try to take over Earth by seamlessly transforming into physical copies of real humans. But the underlying themes of the film express Cold War paranoia about communism, conformity, and mind control.

*Invasion of the Body Snatchers* has been reinterpreted for new audiences a few times since the release of the original, with each iteration highlighting a different set of cultural anxieties. A critically acclaimed 1978 remake, for instance, is more focused on critiques of consumerism than communism.

*The World's End*, a delightful comedy released in 2013, is essentially another *Body Snatchers* remake in disguise as a midlife coming-of-age story. The protagonist, Gary King, is an alcoholic man-child on the cusp of middle age who decides to assemble his old high school friends for a beer crawl. His plans devolve into chaos after it becomes clear their hometown has been invaded by robotic body-snatching aliens. In this iteration, the invasion is thwarted by the utter gormlessness of a man who turns out to be humanity's best weapon, and unconventional representative, against a conformist alien federation.

Some aliens don't quite snatch bodies but instead just simulate or repurpose them, like the mirage-like "guests" of *Solaris*. Unlike body snatchers, these shapeshifters often have ambiguous intentions. Occasionally, they are benevolent; in the 1985 novel *Contact*, cowritten by Carl Sagan and adapted into a movie in 1997, a human crew meets intelligent aliens who take the form of their loved ones to foster a sense of familiarity.

The 1984 film *Starman* offers an especially bizarre alien shapeshifter who takes the form of a widow's dead husband, leading to a

boxcar sex scene and Jeff Bridges's fantastic deadpan delivery of the line "I gave you a baby tonight." Meanwhile, *Avatar* turns the trope on its head by casting a human as the shapeshifter that infiltrates an alien society by disguising himself as one of them.

I could go on, and would, but the Sun might explode before I am done. The upshot is that these shapeshifter aliens are a particularly powerful force in science fiction that plays on the suspicion that there is something alien hiding within the people we encounter in our daily lives. You simply don't get this eerie sensation with openly alien humanoids; Worf is not trying to convince you that he is actually a human just because he looks exactly like one up to his eyeline.

But the possibility that the world is filled with secret aliens, perhaps even in your own social and familial circle, has deep resonance. It can be a comfort when you feel like the only sane person in an insane world. It can be a warning about deceiving outward appearances. And it's always a reminder that no amount of empathy and communication can actually put you in the perspective and mindset of another human being; sometimes the distances between our experiences as people are about as untraversable as interstellar space.

▲ The alien in *Starman* shows off one of his fancy miracle spheres.

## GALACTIC GEOPOLITICS

Human-alien bureaucracies are quite common in science fiction. After all, forging interstellar partnerships with like-minded civilizations seems all well and good until you think about all the paperwork.

Star Trek's United Federation of Planets, a democratic republic of participating sovereign worlds modeled on the United Nations, is an expression of the franchise's general optimism about diplomatic solutions to conflicts—but it's definitely an unwieldy bureaucracy. The Galactic Senate in the Star Wars universe occupies a similar role, while the 1997 comedy *Men in Black* is a portrait of a kind of administrative deep state for handling alien visitors.

But perhaps the ultimate alien bureaucrats are the Vogons in Douglas Adams's *Hitchhiker's Guide to the Galaxy*, who are such natural pencil pushers that, to quote the book, "they wouldn't even lift a finger to save their own grandmothers from the Ravenous Bugblatter Beast of Traal without orders signed in triplicate, sent in, sent back, queried, lost, found, subjected to public inquiry, lost again, and finally buried in soft peat for three months and recycled as firelighters."

## TIME-DISTORTING ALIENS

The reality we experience as humans is pinned in place by the arrow of time; we can only experience the past as memory and the future as conjecture. Escaping the constraints of linear time is one of our essential human obsessions, and it has found exciting manifestations in alien fiction.

Agent J and Agent K enforce galactic bureaucracy in *Men in Black*. ▲

Ted Chiang's 1998 novella *Story of Your Life*, which was adapted into the 2016 film *Arrival*, is a masterful example of this trope. The protagonist, a linguist named Louise, is tasked with establishing communications with aliens, called heptapods, that inexplicably arrive on Earth one day. By gaining fluency in the heptapod language, Louise also mentally absorbs their nonlinear experience of time, in which past, present, and future occur simultaneously. As a result, she experiences moments in her future as if they are memories of the past. The alien-derived foresight in Chiang's story is a linguistic projection of nonlinear time; Louise could only experience the future as a fleeting recollection, rather than a real place.

Hiroshi Sakurazaka's 2004 novel *All You Need Is Kill*, adapted into the 2014 film *Edge of Tomorrow*, explores a similar concept under a very different tonal lens. In this story, Earth is invaded by an alien species called Mimics who can send messages back in time. This ability gives the Mimics a huge competitive advantage; if they die in battle, they can inform their past selves to avoid the exact hazards that killed them. The war doesn't end until a human soldier finds himself locked into one of these time loops, providing the inside edge that humans need against the invaders.

Time loops also serve as the basic story structure of the 2019 video game *Outer Wilds*, in which you play an alien astronaut who must piece together clues about another alien species, called the Nomai, that occupied your native solar system long ago before going extinct. The Nomai's advanced technologies have created a twenty-two-minute time loop that ends with the explosion of the system's star—and therefore your character's repeated death. While both *Edge of Tomorrow* and *Outer Wilds* present an alien species that can manipulate time, they explore the trope with near-opposite tones and outcomes: The Mimics must be annihilated to stop the loop, but empathy for and understanding of the Nomai are the keys out of the temporal cage presented in *Outer Wilds*.

# Exodinosaurs

DINOSAURS WERE EARTHLINGS. WE FIND THEIR remains here on Earth. Given that dinosaurs reigned for so much longer than humans—and still survive today as a myriad of enterprising birds—they were arguably even more Earthy than we are.

So why do we want dinosaurs to be aliens? Or is it that we want aliens to be dinosaurs?

We can learn so much about fictional aliens writ large from the example of the "exodinosaur," a term for "saurian" aliens coined by Allen A. Debus in his book *Dinosaurs in Fantastic Fiction*. Exodinosaurs are not always literal dinosaurs, but they must have typical "saurian" features that are analogous to the humanoid aliens that are ubiquitous in science fiction.

Exodinosaurs mostly exist today as a playful visual trope in science fiction and children's literature. After all, what could be more imaginatively stimulating than placing the most epic avatars of deep time into the immense expanse of deep space?

It was in this spirit that I first encountered exodinosaurs as a kid who compulsively read and reread volumes of Bill Watterson's beloved comic *Calvin and Hobbes*. One of my favorite strips, published on Halloween 1990, features Calvin griping about his teacher's lack of appreciation for his drawing, which he describes as "a beautiful work of power and depth." Hobbes responds, "It's a Stegosaurus in a rocket ship, right?"

By the time the "Stegosaurus in a rocket ship" strip was published, the childlike wonder of exodinosaurs had already entered the cultural bloodstream. But the trope has much earlier origins, and far more profound repercussions, than its apparently trivial nature might imply.

Humans have been discovering the remains of dinosaurs (among other fossils) for thousands of years. It wasn't until the nineteenth century, however, that the concept of a vast extinct lineage of animals, collectively known as dinosaurs, crystallized in the scientific community—and then proceeded to blow the minds of spectators around the world.

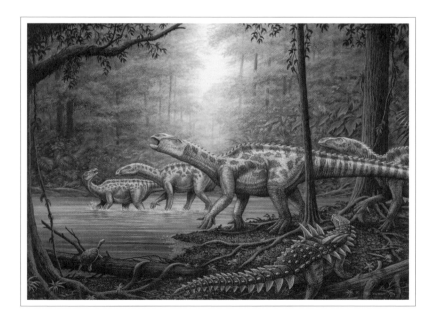

Whereas Copernicus dethroned Earth (and humans) from a privileged spot in space, dinosaurs were part of a wave of findings that revoked our privileged spot in time. This revelation not only upended the conventional timeline of our world, it exposed a new and frightening concept: extinction.

▲ These dinosaurs are probably pondering interstellar travel.

> "People were experiencing a fundamental destabilization of human importance [in the late nineteenth century], of the guarantee of our own survival, and of God's benevolence, now that we learned that He could just wipe out creation. It wasn't exactly reassuring."
>
> —ZOË LESCAZE, *writer and paleoart expert*

In the midst of this existential crisis, dinosaurs began to show up in novels about lost worlds and undiscovered frontiers, which had become a booming genre in the late nineteenth century. On some level, people wanted to believe that a surviving dinosaurian lineage would appear around the next corner. Jules Verne depicted dinosaurs deep underground in *Journey to the Center of the Earth* (1864), while Arthur Conan Doyle's *The Lost World* (1912) placed them on jungle plateaus. Dinosaurs even show up in Antarctica in Frank Mackenzie Saville's *Beyond the Great South Wall* (1899).

With that in mind, it was only a matter of time before dinosaurs eventually ended up in outer space, in Gustavus W. Pope's *Journey to Venus* (1895), an adventure tale about a US Navy officer and a Martian princess who travel to the titular planet.

"Along the shores, basking on the sands or wallowing in the miry fens, were huge land reptiles, the Iguanodons, Megalosaurs, and Dinosaurs," Pope wrote in the book. "These were but the beginnings of the wonders displayed on this Primeval World."

Pope's passage here is the Genesis 1:1 of the exodinosaur bible. Note that Pope imagines Venus as a "Primeval World," a land where the past is somehow still present. This germinal exodinosaurian idea suggests that the lost worlds of our deep past can be recovered by traversing deep space. Venus became the most popular setting in part because of real speculation that it might host dinosaurs, or megafauna like them.

The dream of finding dinosaurs on Venus pretty much died once we sent space probes to our inhospitable neighbor. But crucially, the loss of a Venusian dinosaur homeland only displaced exodinosaurs into a new and more exotic interplanetary diaspora, where they served increasingly complex narrative purposes.

During the early twentieth century, exodinosaurs were often sluggish creatures planted into wild landscapes where they could be conquered by humans to satisfy a feverish demand for swashbuckling frontier fantasies. But by the middle of the century, exodinosaurs began to be tamed, captured, and even understood as sentient minds in their own right.

Take Anne McCaffrey's 1980s Dinosaur Planet book series, set on the fictional planet Ireta, which is populated by intelligent dinosaurian aliens descended from imported zoo animals from Earth. In Donald Glut's 1976 novel *Spawn*, the interstellar trade in exodinosaurs runs in the opposite direction. The novel follows a mission to a distant planet called Erigon to extract alien dinosaur eggs for transport back to Earth so they can be hatched and displayed in a preserve called Dino-World.

Captive dinosaurs also show up in space stations during the 1970s and 1980s, a shift that coincides with the appearance of the first orbital stations launched by the USSR and the US. In George R.R. Martin's novella *The Plague Star* (part of the 1986 collection *Tuf Voyaging*), salvagers access an abandoned seedship called the *Ark* that houses the deadliest creatures in the known universe. Earth's contribution is the *Tyrannosaurus rex*, of course.

Orbital dinosaurs also populate Robert Silverberg's 1980 short story "Our Lady of the Sauropods." The story takes place on Dino Island, a space station filled with genetically engineered dinosaurs developed as commercial attractions that are loosed when the ship's control system goes dark (yes, it's basically *Jurassic Park* in space).

"We know just a shred of what the Mesozoic was really like. Just a slice, literally the bare bones," the narrator says. "The passage of a hundred million years can obliterate all traces of civilization. Suppose they had language, poetry, mythology, philosophy? Love, dreams, aspirations?"

The novel concept of intelligent dinosaurs inspired many popular thought experiments in the 1980s and 1990s. In a universe where the

## An Essential E.T. Playlist

**MOST OF THIS CHAPTER** has focused on aliens in film and in print, but extraterrestrials have also been invading popular music for many decades. To close out, here's a roundup of some of the best bangers about aliens.

"Two Little Men in a Flying Saucer" by Ella Fitzgerald (1951)

"Let There Be More Light" by Pink Floyd (1968)

"Starman" by David Bowie (1972)

"I've Seen the Saucers" by Elton John (1974)

"Unfunky UFO" by Parliament (1975)

"Rapture" by Blondie (1980)

"E.T. (Extraterrestrial)" by OutKast (1996)

"Spaceman" by Bif Naked (1998)

"Aliens Exist" by Blink 182 (1999)

"Rosetta Stoned" by Tool (2006)

"Alien" by Britney Spears (2013)

"ALIEN SUPERSTAR" by Beyoncé (2022)

asteroid missed, for example, could dinosaurs have become a technological civilization that was on par with, or perhaps even beyond, our own human civilization?

*Dinosaucers*, a resplendent animated show which ran for one season in 1987, answered this question with anthropomorphic dinosaurs who "dinovolve" into their ancestral forms and who drive spaceships that are, you guessed it, also shaped like dinosaurs. The Dinosaucers are from a planet called Reptilon that orbits the Sun on the opposite side of Earth and that was therefore not hit by an asteroid 66 million years ago.

This basic idea is explored again in the 1997 *Star Trek: Voyager* episode "Distant Origin" that introduces saurian aliens named the Voth that turn out to be the descendants of intelligent hadrosaurs that left Earth in the fallout of the asteroid strike 66 million years ago.

By imagining dinosaurs elsewhere in the universe, it's as if we are rooting for evolution to follow a similar trajectory on other worlds as it has here on Earth. Even scientists have engaged in this wish fulfillment; the chemist Ronald Breslow and astronomers Lisa Kaltenegger and Rebecca Payne have speculated about the possibility of dinosaurs on other planets. (It's also worth mentioning that dinosaur fossils have literally traveled to outer space as part of space missions, and some researchers have suggested that the asteroid impact may have hurled dinosaur remains onto the Moon.)

In this way, exodinosaurs are an expression of our dreams and nightmares. Don't be fooled by their cartoonish appearance: These animals reflect profound revelations about the impermanence of our time on Earth, both as individuals and as a species. Aside from humanoid aliens, they are among the most common visualizations of extraterrestrial life, expressing a hope that what is dearly departed from Earth can one day be found alive again off of it.

|||||||||||||||||||||||||||||||||||||||||||||||||||||||||||||||

The truth is out there, Mulder,
but so are lies.

—DANA SCULLY, *The X-Files*

|||||||||||||||||||||||||||||||||||||||||||||||||||||||||||||||

CHAPTER FOUR

# FLYING SAUCERS

# "It's never aliens."

**THIS PHRASE HAS BECOME AN UNOFFICIAL CATCHPHRASE** for many researchers involved in the search for extraterrestrial life and the reporters who cover the beat. It emerged reflexively in response to the widespread belief that "it"—a strange flash in the sky, a radio source from deep space, a redacted line in a government document—is indeed aliens.

No, answer the Authorities. It is not, nor has it ever been, aliens. But of course, that's exactly what Authorities might say if it *were* aliens (or, at least, this is how such pronouncements are often perceived by those who believe it's *definitely* aliens).

Over the years, I've reported on many spacey things that are unexplained and often a little unsettling. None of what I have experienced, however, has remotely seemed like slam-dunk evidence of extraterrestrial life anywhere in the universe, let alone here on Earth. But many millions of people have reached the opposite conclusion, in one way or another, propelling the rise of ufology: a loose, shapeshifting field devoted to the investigation of unidentified flying objects (UFOs) or, in modern scientific parlance, unidentified anomalous phenomena (UAP).

Ufology is an American Boomer; it was born in Washington State on June 24, 1947. That was the date that a pilot named Kenneth Arnold spotted a series of bizarre objects traveling at incomprehensible speeds while flying a small plane near Mount Rainier.

That incident is now regarded as the spark that ignited modern ufology and popularized the infamous term "flying saucer." But another strange encounter, which unfolded during the same summer of Arnold's sighting, most fully encapsulates the roiling myths, and unseemly truths, at the heart of ufology.

And that, of course, is the story of Roswell.

> # 𝕽𝖔𝖘𝖜𝖊𝖑𝖑 𝕯𝖆𝖎𝖑𝖞 𝕽𝖊𝖈𝖔𝖗𝖉
> ## RAAF Captures Flying Saucer On Ranch in Roswell Region

# Roswell, Part 1:
## THE FLYING DISC

ON JUNE 14, 1947, A RANCHER NAMED W. W. "Mac" Brazel and his son, Vernon, were driving across their acreage in New Mexico. As the pair surveyed their property, they glimpsed something strange glinting in the desert brush. Upon closer look, it turned out to be the scattered remains of a crashed machine.

Puzzled, the Brazels initially left the wreckage to fester in the summer heat. But weeks later, Mac caught wind of the strange "flying saucers" that people, including the pilot Arnold, kept spotting in the sky. The spate of sightings had snowballed into a public sensation. Was it all a coincidence, he wondered, or was there some connection?

▲ The headline that launched a thousand conspiracy theories.

To get to the bottom of it, Brazel alerted a local sheriff to the existence of the fallen craft on July 6. The next day, military officials from Roswell Army Air Field (RAAF) were dispatched to Brazel's ranch, which was about 80 miles from the New Mexican town of Roswell, to collect the debris. Cool heads seemed to be prevailing—until an RAAF press officer made the fateful decision to describe part of the salvage as "a flying disc" in a press release.

News services around the world rushed to report on the latest flying saucer—which had, this time, left behind physical evidence—while a saucer-hungry public gobbled up the story. In response to the intense interest, an Air Force lieutenant general named Roger Ramey conducted an immediate investigation of the recovered debris.

On July 9, his team announced that the craft was a downed weather balloon, a finding that inspired mockery in the press.

As the Washington, DC, *Evening Star* reported, "An apparently authentic account from an AAF base at Roswell Field, NM, of the recovery of a mysterious aerial disc created international excitement for several hours yesterday until the object was identified as a harmless high-altitude weather balloon."

A completely harmless weather balloon. Much ado about nothing. Or so the government said....

▲ A map of reported UFO crash sites across New Mexico, with the original July 1947 site labeled "Debris Field."

# UFOs Before Ufology

> "It's not a work of man or ghost. What is it?"
>
> —SU SHI, *eleventh-century polymath*

INTUITIVELY, IT MAKES SENSE THAT UFO SIGHTings have proliferated alongside advances in aviation and spaceflight. Humans simply didn't used to toss strange shiny stuff into the heavens all the time. It's no surprise that many of these flying machines are not easily identifiable to observers on the ground.

That said, people have reported seeing strange events in the skies for thousands of years, long before the advent of planes, copters, and rockets. Many historical accounts of these sightings seem to line up with natural processes that have since been explained, but a lot of them don't. UFO observers then, as now, had to rely on their wits and intuition to make sense of what they saw.

## A UFO CRAZE IN THE SONG DYNASTY

About 1,000 years ago, a mysterious object appeared again and again, over a period of at least ten years, in the skies above the eastern banks of the Yangtze River. Eyewitnesses were awestruck by this strange vision and rumors of its mysterious arrivals attracted the attention of the contemporary polymath Shen Kuo. Shen included reports of this skyward "pearl," as he described it in his masterpiece *Dream Pool Essays*, which was completed in the year 1088. The following account, for instance, is eerily reminiscent of modern UFO lore.

"A friend of mine had a retreat by the lake," Shen wrote. "One night he noticed that the 'pearl' was very nearby. At first, it opened its door just slightly. A bright light emerged from its 'shell,' like a single ray of golden thread. A moment later the 'shell' suddenly opened to about the size of half a mat. Inside there was a white light like silver. The 'pearl' was as big as a fist, so bright one could not look at it directly.... Quickly the 'pearl' sped far off, as if it were flying, floating above the waves, bright and brilliant as sunlight."

## FOO FIGHTERS OVER MEDIEVAL NUREMBERG

As dawn broke on April 14, 1561, early risers in Nuremberg stared in wonder at a celestial show unlike anything they had ever seen. It began with a "dreadful apparition" that formed inside the rising Sun, then rapidly escalated into what looked like a violent conflict between warring shapes, according to a contemporaneous report. The event lasted for over an hour and concluded when the shapes seemed to go up in an "immense smoke" and were replaced by a vision of a large black spear.

"Whatever such signs mean, God alone knows," noted that same news broadsheet, which was visually captured in a spectacular woodcut by the artist Hans Glaser.

*Chapter Four:* FLYING SAUCERS ▶ 089

**FOO FIGHTER:** A phrase for unidentified aerial phenomena that predates flying saucers, UFOs, and UAP. The nonsense word *foo* originated in Smokey Stovey cartoons and was picked up by Allied aircraft pilots during World War II to describe the strange aerial phenomena they experienced. Generally, the pilots added another alliterative expletive to their reports of these events.

▲ Hans Glaser's woodcut interpreted the 1561 event as a warning from God and begged Him to "avert his wrath."

## "Flying Saucer": An Unlikely History

**THE TERM** *flying saucer* evokes a classic UFO image of a disc-shaped alien craft, sometimes with extra flourishes like a tractor beam or panoramic windows. But the real history of this iconic phrase is muddier. It first became widespread after the pilot Kenneth Arnold described the UFOs he had witnessed near Mount Rainier as moving "like a saucer if you skip it over water." Like a game of ufological telephone, the local *East Oregonian* newspaper reported that Arnold had seen a "saucer-like" craft, which turned into "supersonic flying saucers" once publications with wider reach picked up the story.

In this way, Arnold's use of the word *saucer* as a descriptor of movement morphed into a descriptor of shape, in what "may be the greatest misquote in the history of journalism," as Adam Frank writes in *The Little Book of Aliens*.

Plenty of disc-shaped UFOs had been reported or depicted prior to the twentieth century, but the modern iconic image of these "flying saucers" is basically due to a reportorial accident.

"If Arnold hadn't said a word history probably would have nevertheless been set on a similar course," Sarah Scoles writes in her book *They Are Already Here: UFO Culture and Why We See Saucers*. "Perhaps, in a world without Arnold's encounter, people would have described 'The Phenomenon' differently. Perhaps we wouldn't have the term "flying saucer" at all. Maybe it would have been pancakes or spheres. But *Arnold* and *saucer* are what we've got."

Remarkably, Arnold's account wasn't even the first time the meaning of *saucer* had been garbled as part of a UFO story. In January 1878, a farmer named John Martin was out hunting near Denison, Texas, when he saw a dark object flying to his south. Martin used the word *saucer* to describe the apparent size of this craft, not its shape, in an article published the same month in the *Denison Daily News* on January 25, revealing that the word *saucer* has a particularly slippery history.

▶ The so-called "Passaic UFO photos," named after the New Jersey town where they were captured in July 1952 by photographer George Stock, are among the most iconic flying saucer images in history.

# "What Did You Actually See?"
## A UFO EXPLANATION CHECKLIST

**T**HE TERMS *UFO* AND *UAP* CONJURE UP IMAGES of flying saucers and little green men, but these acronyms just mean that an observer saw something in the sky that they could not explain. Maybe you did really see an alien spaceship that one time around the campfire or driving back from work. But for researchers who try to identify these curious sightings, there are a few common phenomena that can solve the mystery without invoking E.T.

### EXPLANATION: IT WAS MOTHER NATURE

- **METEORS AND METEORITES:** You saw some rocks burning up in the atmosphere.

- **COMETS:** You saw some shiny ice balls in space.

- **VENUS:** You saw the planet Venus, which can, at times, appear hypnotically bright and strange.

::: **SUN DOGS:** You saw weird bright patches of light that appeared at the same angle on each side of the Sun. These halos are created when sunlight bounces off ice crystals in the atmosphere.

::: **LENTICULAR CLOUDS:** You saw stationary globs of water vapor that form over mountain ranges. Lenticular clouds can take on spectacular forms, including classic flying saucers or towering pagodas, as a result of the alpine topography below.

::: **FATA MORGANA:** You saw an optical illusion right above the horizon caused by a refracted distortion of a real object.

> **THE MUTUAL UFO NETWORK (MUFON),** founded in 1969, is one of the largest nonprofits devoted to civilian volunteer investigations of UFO sightings, with several thousand members worldwide.

▲ Lenticular clouds are nature's flying saucers.

## EXPLANATION: IT WAS HUMAN NATURE

- **FLARES AND FIREWORKS:** You saw a pyrotechnic display, like a distress signal or a celebratory event.

- **SEARCHLIGHTS:** You saw a beam of light supercharged by an artificial parabolic reflector.

- **BALLOONS:** You saw one of the thousands of research balloons that are floating through the sky at any given time.

- **AIRCRAFT:** You saw one of the thousands of aircraft that are soaring through the sky at any given time.

- **SATELLITES:** You saw one of the thousands of spacecraft that are sailing through space at any given time.

- **SKY LANTERN:** You saw a small paper balloon with a flame at its center floating through the air. Sky lanterns are typically launched during festivities and have been the source of many UFO reports.

- **LENS FLARE:** You saw a bright patch of scattered light through your binoculars, telescope, or camera.

- **HOAXES:** You saw a real object that was fraudulently designed to look like an alien spacecraft or alien body.

- **HALLUCINATIONS:** What you saw was all in your head.

▲ Of the thousands of balloons in the sky right now, only a handful are likely to be made by aliens. (I kid!)

# Notable Hoaxes

**P**RANKSTERS AND UFOS ARE A MATCH MADE IN heaven. What better way to prove your mastery of mischief than to mislead people into believing they have glimpsed, with their own eyes, something not of this world? For over a century, a variety of colorful charlatans and tricksters have successfully engineered fake UFO sightings or sensational footage that have profoundly shaped alien folklore—even if the events were later revealed to be hoaxes. Here are a few of the most notable instances of UFO tomfoolery.

### THE DUNDY COUNTY HOAX

On June 6, 1884, a group of cowboys witnessed a piece of airborne machinery fall from the skies and crash into the prairies of Dundy County, Nebraska. At least, that's what the *Daily Nebraska State Journal* reported the following morning, calling the ship a "celestial visitor" that was "evidently a machine of human manufacture." But the tale quickly grew extraterrestrial legs as more embellishments were heaped upon it by other reports that the craft came from another planet and dissolved completely upon contact with water.

The author of these reports, J. D. Calhoun, was known for spreading tall tales; in 1927, his assistant revealed the entire Dundy County UFO story was fabricated. But though it was fraudulent, the story eerily anticipated a series of mysterious airship sightings that occurred across the US more than a decade later, around the turn of the twentieth century, including a famous report of an alien pilot who crashed near Aurora, Texas, in 1897. Some ufology researchers consider these airship sightings to be the first documented UFO flap.

> **FLAP:** The term for a burst of similar UFO sightings in a small region over a short period.

### THE "OTHER ROSWELL"

During the 1940s, a pair of prolific con artists named Silas M. Newton and Leo A. Gebauer were hawking so-called doodlebugs—devices they claimed could detect valuable resources like oil or gold—to any mark they could find. When UFO-mania gripped the nation during the summer of 1947, Newton and Gebauer saw an opportunity to ride the craze and signal-boost the hustle in the process.

Gebauer convinced the reporter Frank Scully that he was one of an elite group of scientists selected by the US Air Force to view an alien spaceship that had crashed in a canyon near Aztec, New Mexico, in March 1948. Using the alias "Dr. Gee," Gebauer told Scully that the ship was about 100 feet wide and contained the charred bodies of sixteen humanoid aliens who were about 3 feet tall and wore matching blue uniforms.

Gebauer claimed that two other saucers crashed in Arizona following the first and that he witnessed a

▲ Author Frank Scully (right) talks to con artist Silas Newton (center) about their totally above-board and verifiable claims of alien-derived technologies.

fourth spaceship, with living aliens, that vanished upon discovery. He concluded that the extraterrestrials were from Venus. Newton, who had fashioned himself as a brilliant oil entrepreneur, spread news of the crashes in a lecture in the spring of 1950 under the alias "Mr. X," the same year that Scully published the pair's account in his bestselling book, *Behind the Flying Saucers*. As the tale circulated, Newton and Gebauer conned several investors into buying doodlebugs that they said contained alien technology.

In 1952, the investigative reporter J. P. Cahn outed the tale as a hoax in a lengthy article in *True* magazine, writing that Scully's "loudly bad book" had popularized "one of the greatest scientific hoaxes to hit the country." Newton and Gebauer were convicted of fraud a year later. But like so many hoaxes, the Aztec saucer crash still left a major imprint on ufological culture, especially by pioneering the trope of dead alien bodies.

## THE MORRISTOWN UFO HOAX

On the evening of Monday, January 5, 2009, people across Morris County, New Jersey, witnessed mysterious red lights floating in the sky, fueling a wave of police reports and speculation about extraterrestrial visitors.

The mystery and intrigue only deepened after the lights appeared on four other nights over the course of the next six weeks, often seeming to fly in formation. The final appearance of the lights, on February 17, was so dramatic that government authorities in the region expressed concern that the UAP, whatever it was, might interfere with aircraft, including passenger jets descending into Newark Liberty International Airport.

While the event attracted media coverage and input from UFO experts, the origin of the lights as a hoax was not confirmed until, of course, April Fool's Day, when Joe Rudy and Chris Russo copped to creating the lights by tying flares to helium balloons as part of an "experiment" in an article for *Skeptic* magazine.

"We brainstormed the idea of producing a spaceship hoax to fool people, bring the charlatans out of the woodwork to drum up controversy, and then expose it as nothing more than a prank to show everyone how unreliable eyewitness accounts are, along with investigators of UFOs," they said.

"We delivered what every perfect UFO case has: great video and pictures, 'credible' eyewitnesses (doctors and pilots), and professional investigators convinced that something amazing was witnessed. Does this bring into question the validity of every other UFO case? We believe it does."

Rudy and Russo even participated in the flurry of media attention by sharing their footage of the lights at local conventions and on the radio. While they clearly had fun with their experiment at the time and conducted it for what they believed was the public's benefit, they were ultimately reprimanded and charged with disorderly conduct, receiving a sentence of a small fine and community service.

## CROP CIRCLES

Few symbols are more emblematic of the belief in alien visitation than crop circles, which are made by flattening tall crops, like wheat or corn, into large-scale patterns across a field. These agricultural designs have been observed for more than a century, but their association with aliens and UFOs began to click in the 1960s, after geometric designs were reported in locations everywhere from rural Alberta to northern Australia. Many locals connected crop circles with UFO sightings, creating a suspicion that extraterrestrial activity was behind the strange patterns.

During the 1970s, a massive proliferation of crop circles swept across sleepy English cornfields, cementing the phenomenon as a public sensation. Pranksters Doug Bower and Dave Chorley later took credit for more than a hundred of those English crop circles and revealed that several independent groups had participated in the hoaxes across the nation.

There are still many people who believe that at least some crop circles have an extraterrestrial origin, but the general consensus is that these designs are human-made, either as hoaxes or intentional artwork.

## CATTLE MUTILATION

For decades, farmers and ranchers have discovered carcasses of livestock with bizarre pathologies, such as missing tongues or broken legs, fueling gruesome speculation about alien visitations.

These theories were inflamed by waves of livestock mutilations in the American Southwest during the 1970s, which prompted several investigations, including an effort headed by former FBI agent Kenneth Rommel. In a 1980 letter to the FBI, Rommel noted that "certain segments of the population have attributed the damage to other causes ranging from UFOs to a giant government conspiracy" but concluded that the damage was attributable to "normal predator and scavenger activity."

Reports of livestock mutilations have persisted to the present day. Though they are still typically caused by predators or animal abuse by humans, many people maintain suspicions of a supernatural or extraterrestrial cause.

▲ A typical geometric crop circle in Switzerland.

UNCLASSIFIED

Technical Report  No. F-TR-2274-IA

# UNIDENTIFIED AERIAL OBJECTS
## PROJECT "SIGN"

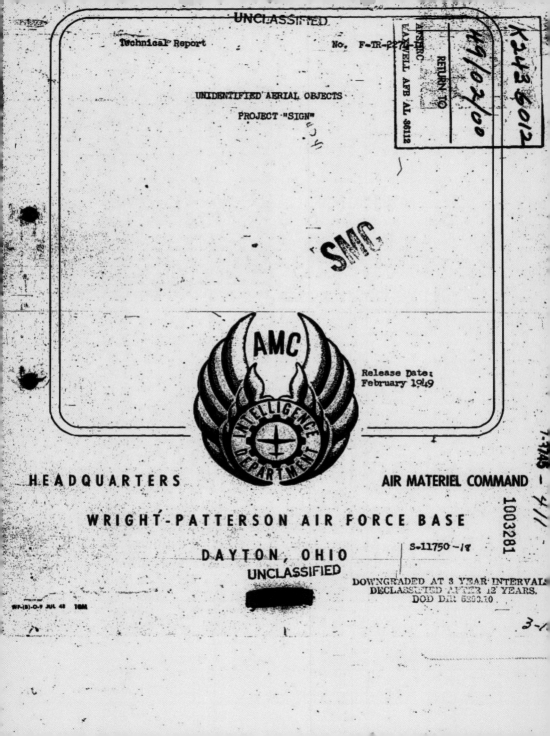

Release Date: February 1949

**HEADQUARTERS**     **AIR MATERIEL COMMAND**

**WRIGHT-PATTERSON AIR FORCE BASE**

**DAYTON, OHIO**

UNCLASSIFIED

DOWNGRADED AT 3 YEAR INTERVALS
DECLASSIFIED AFTER 12 YEARS.
DOD DIR 5200.10

# TOP SECRET: Governments React to UFOs

As UFO mania began to spread around the world in the mid-twentieth century, governments launched investigations and filed reports to address civilian sightings. But governments were not passive responders; the shady protocols and dismissive attitudes of some early investigations bred distrust between federal officials and the ufological community. As a consequence, the "government"—especially in the United States—is often cast as a primary villain in ufological lore. Here are some of the most important government investigations into UFOs, most of which were once Top Secret and have since been declassified.

::: **PROJECT SIGN (NÉE SAUCER):** Launched in 1948, Project Sign was the US government's first coordinated response to the UFO craze that had erupted the previous year. Initially named Project Saucer, the investigation concluded that most UFO sightings could be linked to known natural and technological events, but that some reports defied explanation. For that reason, the Sign team recommended that the government continue studying UFO sightings. There was much more work to be done.

◄ The cover of the now-declassified Project Sign, the first US investigation into UFOs.

- **PROJECT GRUDGE:** The successor to Project Sign at the US Air Force, Project Grudge earned its ominous name, in some ways. Conducted in 1949, Grudge was motivated by fears within the US government that American UFO believers could be vulnerable to psyops and other forms of mind control or manipulation from rival nations, especially the Soviet Union. To that end, the Grudge report, released in the summer of 1949, attempted to calm the public by downplaying the risks of UFO sightings. The report chalked the sightings up to factors such as misinterpretation, hoaxes, "mass hysteria and war nerves," and "psychopathological persons."

- **PROJECT BLUE BOOK:** This eighteen-year effort launched in March 1952 catalogued and investigated 12,618 sightings reported between 1947 and 1969; 701 of these events remain unexplained. In their final report, the project's investigators concluded that none of the evidence submitted to the Air Force represented "technological developments or principles beyond the range of present day scientific knowledge" and that "there has been no evidence indicating that sightings categorized as unidentified are extraterrestrial vehicles."

- **THE CONDON REPORT:** A US Air Force committee tasked with examining results from Project Blue Book, and headed by the physicist Edward Condon, issued this 1968 document. Condon and his colleagues compiled their findings into a 965-page report that concluded "nothing has come from the study of UFOs in the past 21 years that has added to scientific knowledge" and that "further extensive study of UFOs probably cannot be justified in the expectation that science will be advanced thereby."

  The report had a major chilling effect on official academic and governmental research into UFOs for several decades and delivered a killing blow to Project Blue Book.

## UNCLASSIFIED
## ~~CONFIDENTIAL~~

### U. S. AIR FORCE TECHNICAL INFORMATION SHEET

This questionnaire has been prepared so that you can give the U. S. Air Force as much information as possible concerning the unidentified aerial phenomenon that you have observed. Please try to answer as many questions as you possibly can. The information that you give will be used for research purposes, and will be regarded as confidential material. Your name will not be used in connection with any statements, conclusions, or publications without your permission. We request this personal information so that, if it is deemed necessary, we may contact you for further details.

1. When did you see the object?

   Day ____ Month ____ Year ____

2. Time of day: ____ Hour ____ Minutes

   (Circle One):  A.M.  or  P.M.

3. Time zone:
   (Circle One): a. Eastern
                 b. Central
                 c. Mountain
                 d. Pacific
                 e. Other ____

   (Circle One): a. Daylight Saving
                 b. Standard

4. Where were you when you saw the object?

---

- **FLYING SAUCER WORKING PARTY:** This magnificently named investigation, commissioned in 1950, was the United Kingdom's first official study of UFOs within British borders. The FSWP issued a report in 1951 recommending that "no further investigation of reported mysterious aerial phenomena should be undertaken" unless and until some material evidence should become available.

- **PROJECT MAGNET:** Canada's governmental transportation agency, Transport Canada, launched Project Magnet in 1950 with the aim of understanding UFO sightings within the nation's vast borders. The group was headed by the engineer William Brockhouse Smith, who ultimately concluded that aliens were visiting Earth, though his view was not supported by his colleagues.

▲ The Air Force gathered data for Project Blue Book by providing questionnaires to people who reported seeing UFOs.

# Roswell, Part 2:
## AN INSIDE MAN

**E**VER WONDERED WHAT IT MIGHT BE LIKE TO nuke some battleships? That's exactly what the US military was pondering in the fallout of World War II. To find out, the Air Force dropped two atomic bombs on dummy war fleets near Bikini Atoll, a remote tropical island in the Pacific Ocean, in 1946. A bomb named Gilda was dropped on July 1 and exploded in midair. Another bomb named Helen followed weeks later, on July 25, and blew up underwater.

Jesse Marcel, an Air Force major, was there to help administer the tests, known as Operation Crossroads. He witnessed the surreal scene of mushroom clouds erupting out of a paradisiacal island lagoon. It's hard to imagine anything more memorable than this terrible and awesome sight. But Marcel came to be known for something else he saw—or claimed to have seen—the following summer, near Roswell, New Mexico.

Marcel was one of the military officers who drove out to the Brazel ranch on July 7, 1947. He helped gather the remaining debris, then loaded it into his Jeep and delivered it to the Roswell base. The next day, Marcel escorted the aircraft's remains by plane from Roswell to Fort Worth Army Air Field in Texas, where a press conference had been convened. At the event, Marcel confirmed that the crashed debris was a "weather device" and was pictured with some of the salvage.

Over the next few years, Marcel was promoted to lieutenant colonel and transferred to a few more posts before moving his family back home to coastal Louisiana to care for his mother. He entered civilian life and faded into obscurity for decades.

Then, in 1978, Marcel suddenly reemerged—and he had a secret to tell. In a serendipitous interview with Stanton Friedman, a colorful ufologist, Marcel revealed that the wreckage he had recovered from the Brazel ranch was not a weather balloon. It was an alien spaceship. The government had lied and executed a cover-up.

Marcel's account began to circulate in tabloids and took center stage in the 1980 book *The Roswell Incident* by UFO researchers Charles Berlitz and William Moore. In addition to popularizing the Roswell story, Berlitz and Moore embellished it with claims that actual alien bodies, in addition to technologies, had been recovered at the Brazel farm, a detail that was influenced by the 1948 Aztec saucer hoax, among other rumors.

For his part, Marcel never said there were bodies, but the story no longer belonged to him—or anyone. It had become like the Blob, swallowing anything it came across, gaining strength and momentum by incorporating existing unaffiliated tales into its ufological bulk and evolving entirely new lore.

In fact, over the course of the 1980s and 1990s, ufologists injected so much narrative lifeblood into the Roswell myth that the community experienced a series of internal schisms over the exact details of the initial crash and the alleged government cover-up.

▲ Air Force Major Jesse Marcel poses with the wreckage recovered near the 1947 crash site at Roswell, New Mexico.

"Enthusiasts are now even charging that for 40 years the Federal Government has harbored evidence of an encounter with extraterrestrial creatures, including their lifeless bodies and damaged spacecraft," reported the *New York Times* in 1987. "That startling report, dismissed by skeptics and Government officials as laughable, is contained in what purport to be top-secret Government papers from the Eisenhower era."

The report referenced in the quote was, in fact, revealed to be a hoax a few years later, a development that further inflamed divisions within the UFO community. These internecine conflicts, combined with more unhinged accounts, muddied the Roswellian waters and helped to broadly discredit UFO reports to the public.

But tall tales sometimes contain a kernel of truth. And in 1994, the government would finally admit that there was a cover-up at Roswell—though it had much more to do with terrestrial paranoia than anything from beyond Earth.

---

"Many motives have been ascribed to the senders or occupants of UFOs, mostly concerned with attack, exploitation, reconnaissance for conquest, capture of human specimens, or planned future occupation of the earth, all of which mirror our own images of how we would react in their place. But, possibly because of the danger we represent to ourselves and our surroundings, there may be another explanation. Perhaps what we call UFOs are part of a design—or message—whose meaning may become clear to us, one hopes, while there is still time."

—**BILL MOORE AND CHARLES BERLITZ,**
*The Roswell Incident* (1980)

# Barney, Betty, and the Birth of Alien Abductions

**I**T WAS LATE INTO THE EVENING OF SEPTEMBER 19, 1961, and only starlight illuminated US Route 3 through New Hampshire. A married couple named Barney and Betty Hill, along with their dachshund, Delsey, were driving back home from their honeymoon in Quebec. Around 10:30 p.m., Betty, who was in the passenger seat, noticed a strange light in the sky that she presumed was a shooting star. After Barney pulled over to observe it for himself, he also tried to come up with a reasonable explanation: Was it perhaps a commercial jet?

Puzzled, the Hills resumed their journey, while keeping an eye on the strange object that now appeared to be trailing them. At one point, the craft swooped out from behind a face-like cliff structure known as the Old Man in the

▲ Barney and Betty Hill recount their abduction experience while Delsey the dog looks on.

Mountain, then descended rapidly toward the couple's Chevy Bel Air, forcing Barney to stop in the middle of the highway.

What they say happened next remains one of the biggest flash points in ufological history. Barney recalled leaving the car and seeing figures that were humanoid, but not human, inside the craft. He perceived a message from these beings instructing him to remain where he was and keep his eyes on the ship. He was scared. The Hills heard a series of beeps and became disoriented, then regained their senses after a similar sound pattern. They were suddenly back on the road again, alone, miles from where they had stopped.

The Hills reported the experience to a UFO research organization and eventually underwent hypnosis in an attempt to dislodge details about their apparent encounter. The stories that have emerged from those hypnosis sessions, and from the couple's conscious recollections, cemented a powerful class of encounter: the alien abduction.

Betty and Barney recalled arriving at home later than expected, as if they had lost a chunk of time during their journey. They reported a host of eerie sensations and revelations in the days that followed. The

▲ The Hills under hypnosis. (Delsey not pictured.)

clothes they wore that night were damaged or stained with strange substances, their watches stopped working, and the car was etched with circular figures. Betty had vivid dreams and Barney felt an unexplained impulse to examine his genitals, though he found nothing wrong. While hypnotized, the couple reported that the humanoid aliens had brought them into their craft and communicated with them telepathically.

The Hills were not the first people to report an alien abduction; rumors about extraterrestrial kidnappers had circulated since the turn of the twentieth century. Barney and Betty did not seek publicity, but their report wound up exploding into a media sensation a few years later after a journalist shared their experience in a 1965 article, sparking a wave of similar accounts in subsequent decades.

Some skeptics have suggested that the Hills were simply sleep-deprived or swayed by pop culture depictions of aliens at the time, which included tales of abduction by shadowy humanoids. Historians have suggested that racial undercurrents were also baked into the tale: Barney, a Black man, and Betty, a white woman, were both civil rights activists who lived in an era that was actively hostile to interracial relationships, and that may have fostered a sense of paranoia and alienation. While we'll never know the real events of that fateful autumn night, there is no doubt that it had an enormous impact on ufology and pop culture.

Indeed, Barney and Betty have been described as the "Adam and Eve of alien abduction stories" because so many elements of their experience have been repeated in subsequent reports: Missing time. Damaged personal items. Strange memories and dreams. Telepathic communication. A sense of having been touched, and sometimes sexually violated.

Not all abduction reports follow this pattern, but some elements appear as very common threads to this day—though it's worth noting that abduction reports have fallen dramatically since their peak in the 1970s.

## ⋮⋮⋮ Other Important Abduction Tales

⋮ **November 1953:** Riley Martin, a radio host and author, claimed that he was abducted by aliens near the St. Francis River in Arkansas, an event that came to define his whole life. Martin believed he was in contact with an extraterrestrial species known as the Biaviians. He even credited one of them, named O-Qua Tangin Wann, with coauthoring his book *The Coming of Tan* on their worldview.

⋮ **October 15, 1957:** Antônio Villas Boas, a farmer, claimed he was abducted by aliens while plowing fields in the Brazilian state of Minas Gerais. He recalled a sexual encounter with an alien on board the ship, who signaled she would bear his child in space, before he was released, resentful that he had been manipulated into being "a good stallion," in his words.

⋮ **November 5, 1975:** Travis Walton, a forestry worker, claimed he was abducted by aliens while on-site in the Apache-Sitgreaves National Forest in Arizona. Walton was reported missing for several days, and the story became a sensation that inspired the 1993 film *The Fire in the Sky*. However, Walton's claims were undercut by a series of revelations, including the admission by a close friend that the event had been staged.

▲ Most of us don't mind being abducted from time to time, but at least leave our cars alone.

- **May 10, 1978:** Jan Wolski, a farmer, claimed he was abducted by aliens near the Polish village of Emilcin. He reported that the aliens took him and ordered him to undress for an examination in their flying saucer, and offered him food before releasing him. A monument to the alleged abduction now stands where it is said to have occurred, with the phrase "The truth will astonish us in the future."

- **Aliens from Below:** When you hear of UFO sightings or alien abductions, it's natural to picture extraterrestrials arriving from outer space. But there is also a subset of lore that predicts aliens may already be present on Earth deep underground, or hidden in the ocean. This Lovecraftian idea has given rise to real ufological theories, such as the rumor of a crashed alien spaceship that lies under the Bermuda Triangle, as well as a host of fun science fiction about aliens that lie in wait to attack, including the Pacific Rim franchise or *The Tomorrow War*.

**AREA 51:** An extremely top secret, highly classified, don't-even-try-to-get-in US Air Force facility located near Rachel, New Mexico. First established in 1955, Area 51 has been a testing ground for advanced military technologies, such as the U-2 and Blackbird aircraft. The site, which officially goes by the far more innocuous-sounding name Homey Airport, is also a major focal point of ufological conspiracy theories. For decades, rumors have circulated that the government is hiding crashed alien spacecraft, or even alien bodies, at Area 51.

**SKINWALKER RANCH:** A 500-acre ranch near Ballard, Utah, that has been a hotbed of strange anecdotal reports for decades, including UFO sightings, mutilated or disappeared livestock, and crop circles. The ranch became especially prominent when it was acquired by Robert Bigelow, a real estate magnate and UFO believer, in 1996. (He has since sold it.)

# Roswell, Part 3:
## A WHITE LIE

In the years following World War II, the United States not only conducted its own nuclear tests—including those witnessed by Jesse Marcel—it also developed a range of surveillance technologies to figure out if any other countries were nuking stuff elsewhere on Earth.

One of these top secret efforts, Project Mogul, involved launching huge, weird-looking balloons that were designed to detect atmospheric reverberations of nuclear tests within the Soviet Union from cruising altitudes high in the stratosphere. Once fully deployed, the Mogul balloons extended nearly 700 feet in height, and they carried specialized equipment, including radar reflectors made of shiny metal foil.

It was one of these balloons, launched on June 4, 1947, from Alamogordo, New Mexico, that ultimately crash-landed on the Brazel farm northwest of Roswell—at least according to an Air Force report on the "Roswell Incident" released in July 1994.

"Comparison of all information developed or obtained indicated that the material recovered near Roswell was consistent with a balloon device and most likely from one of the Mogul balloons that had not been previously recovered," the report concluded. "Air Force research efforts did not disclose any records of the recovery of any 'alien' bodies or extraterrestrial materials."

Given the highly classified status of Project Mogul, the true nature of the wrecked balloon was not given at Marcel's 1947 press conference about the recovered debris. The craft was instead written off as a weather balloon, which reporter William Broad described as a "white lie" in his coverage of the 1994 event.

"Over the decades, the incident grew to mythic dimensions among flying-saucer cultists," Broad wrote, adding that "Roswell was the greatest of all governmental cover-ups."

Roswell did not turn out to be remotely the greatest of all government cover-ups; it's not comparable to the scale of Watergate, or

▲ An array of Project Mogul balloons from the late 1940s that—supposedly—were the source of the Roswell wreckage.

Iran-Contra, or whatever other unsavory incidents may have been successfully eighty-sixed. But it was, at least, a little itty-bitty cover-up. And that mattered, because it came to fit into a wider tapestry of government deception and subterfuge about UFOs that, in many ways, originated at Roswell.

In 1997, after decades of mixed messages about UFOs, the US Central Intelligence Agency (CIA) decided to cop to some of its past misbehavior in a study called the "CIA's Role in the Study of UFOs. 1947–90." The document revealed that over half of all UFO reports in the late 1950s and 1960s in the United States were identifiably top secret military aircraft, such as the U-2 or Blackbird stealth planes. As with the Mogul balloons, the government lied to the public about the nature of these reconnaissance flights to protect their sensitive status.

The Air Force made "misleading and deceptive statements to the public in order to allay public fears and to protect an extraordinarily sensitive national security project," the study said. "While perhaps justified, this deception added fuel to the later conspiracy theories and the coverup controversy of the 1970s," referring to allegations from the UFO community that the government was hiding information about sightings and encounters with aliens.

Indeed, some people reacted to these reports from the Air Force and the CIA with the same suspicious side-eye as all previous investigations. John E. Pike, who was head of space policy at the Federation of American Scientists in 1997, thought the CIA's admission suggested that government cover-ups of UFOs were probably common.

"The flying-saucer community is definitely onto something," Pike said in a contemporaneous article about the 1997 report.

▲ *The Roswell Report*, the official Air Force report on the alleged 1947 UFO crash, declared the "case closed." Wishful thinking.

## Alien-centric Religions

**IT IS PROFOUNDLY** human to believe that there is a larger meaning to the universe that involves sentient minds beyond Earth (see "Aliens and Religions," page 11). This ancient premonition has found new vectors through ufology and the search for extraterrestrial life, fueling a proliferation of religions based around UFOs, alien visitation, telepathic communication with aliens, and astral projection to places beyond our planet. Examples include:

### SCIENTOLOGY

**A RELIGION FOUNDED** by the author L. Ron Hubbard in the 1950s, which has become the most famous and controversial alien-adjacent spiritual movement to date. Scientology contains space operatic elements in its mythology, including epic battles and evil alien overlords. It is also based in part on the idea that humans have accumulated memories from living many past lives, including in advanced extraterrestrial societies.

### RAËLISM

**A RELIGION FOUNDED** in the 1970s by the French journalist Claude Vorilhon based on the belief that an advanced alien species called the Elohim created humanity with their technologies. Like followers of the Aetherius Society, Raëlian believers think that major religious and historical figures were messengers from beyond Earth that are watching over humanity.

### THE AETHERIUS SOCIETY

**A SOCIETY FOUNDED** in the 1950s by the British author George King, who taught that religious figures in history like Buddha and Jesus Christ were "Cosmic Masters" who originated in outer space. These Cosmic Masters are benevolent guardians who keep humanity safe from evil alien invaders.

▲ The Aetherius Society logo symbolizes the phrase "God manifesting itself as wisdom."

# Twenty–First Century UAP

During the back half of the twentieth century, ufology evolved into a major global phenomenon with distinct narratives and thriving subcultures. The community still retains much of its energy and original mythos, but it has also evolved alongside of the internet and smartphones, which have made it easier to network with like-minded people and share possible sightings.

In addition, perceptions of UAP have shifted considerably as a result of the so-called Pentagon UFO videos, which were released by the Department of Defense (DoD) in 2017. The videos show strange craft with auras and bizarre motions captured by forward-looking infrared (FLIR) targeting instruments on board US Navy fighter jets in 2004, 2014, and 2015. They were collected and examined as part of the DoD's Advanced Aerospace Threat Identification Program, a years-long effort to study UAP.

For instance, two US Navy pilots flying F/A-18F Super Hornets off the coast of San Diego in November 2004 witnessed a strange craft moving in bizarre ways and were able to back up their finding with FLIR recordings. One of the pilots, Commander David Fravor, recalled telling a colleague later that he had no idea what the craft was, noting "it had no plumes, wings, or rotors and outran our F-18s." But naturally, he added, "I want to fly one."

NASA has also convened official teams to investigate bizarre UAP sightings, including repeated reports of spherical metal orbs with no known explanation. The agency's independent study team is meant to signal transparency and remove stigma surrounding the reports of strange objects and events.

This new government posture of interest in and acceptance of the validity of UAP reports may help to single out the most compelling and credible footage and accounts of these inexplicable objects. However, the mutual suspicions that run between government agencies and committed ufologists are still strong. These tensions have, in some ways, been heightened by the government's efforts to appear transparent: Take the 2023 House testimony of David Grusch, a US Air Force officer who claimed that the federal government is trying to reverse engineer technologies from "non-human spacecraft" and that it has recovered mysterious bodies. NASA and the DoD denied the claims, but Grusch's testimony further confirmed the perception of cover-ups among many UFO believers.

The truth may be out there, of course, but the real question for those interested in UAP remains: Whose truth matters?

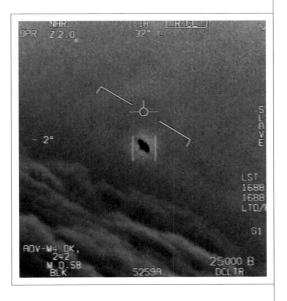

▲ An iconic UAP image captured by US Navy pilots in 2015. In the video footage, one pilot exclaims, "Look at that thing, dude!"

# Roswell, Part 4:
## "WE BELIEVE"

IN THE SUMMER OF 1947, A BALLOON SENT ALOFT to the skies to sniff out a coming apocalypse ended up smashed to bits on a desert ranch instead. That is now the official government explanation for both the debris found near Roswell and the initial deception about its true purpose.

But the tale of Roswell has, ever since, been in a kind of slow-motion cultural crash of its own that has scattered millions of imaginative shards into our minds. It has inspired a slew of books and the equivalent of its own cinematic universe. Kyle MacLachlan, at the height of his first wave of fame, starred as Jesse Marcel in a popular 1994 television movie. In *Independence Day*, government officials confess that the tentacled alien invaders of Earth initially crashed in Roswell. Roswell has had cameos in *The X-Files*, *Futurama*, and *Indiana Jones in the Crystal Skull* and served as the titular setting for the 1999–2002 series *Roswell*, to name a few creative works. The tale has been metabolized into popular folklore and resurfaced in countless different permutations.

There is much that is unreal about Roswell, but it is a real place. Once a quiet desert town known for dairy farms and retirement homes, Roswell has cashed in on the interest and reinvented itself as the world's ufological capital. It is home to the International UFO Museum and Research Center, where Roswellian totems are preserved and

chronicled like religious texts. Thousands of people flock to the town each June for its annual UFO Festival. The municipal motto is "We Believe" in honor of the spacey ufological brand that it has cultivated.

Whatever your actual beliefs about the Roswell incident, there is no denying that part of this culture is simply about colorful kitsch and irreverent speculation. The story embodies the overall trajectory of ufology over the last seventy-odd years, evolving from an early paranoid fever dream to a mass, shared phenomenon to a pop-culture meme. Reports of UFOs have dropped off significantly in recent decades, and the US government has tried to foster a new culture of transparency about these events (at least publicly). But the imprint of Roswell, and its successors, is permanently ingrained, and the truth will forever be "out there."

▲ The city of Roswell, New Mexico, has embraced its alien bonafides.

||||||||||||||||||||||||||||||||||||||||||||||||||||||||||||||||

Our solar system is actually a wild frontier,
teeming with different, diverse places:
planets and moons, millions of objects of ice
and rock.

—CARRIE NUGENT, *planetary scientist*

||||||||||||||||||||||||||||||||||||||||||||||||||||||||||||||||

# CHAPTER FIVE

# THE ALIEN NEXT DOOR

Our solar system has become a kind of universal pictorial language; an iconography of familiar characters that adorns the walls of school classrooms, next to the alphabet and number sets.

**BUT FOR A LONG TIME, HUMANS HAD NO IDEA THAT** we were even in a solar system. When we finally figured it out, it still took us a while to grasp its basic configuration and the immense distance that separates us from other star systems. And we remain in the dark about whether or not it is the only place in the universe that hosts life.

Our evolving view of the solar system reflects the expanding boundaries of our imagination. We noticed the distinct paths of planets against a starry backdrop and were able to project ourselves outward to identify them as our sibling worlds. We invented telescopes that revealed that some worlds, such as Jupiter and Saturn, are mini-systems unto themselves. And now we are peering across interstellar space at thousands of star systems and ascertaining how our neighborhood measures up in the great galactic census.

The most surprising upshot of this journey is that our solar system appears to be an extraordinary outlier as a harborer of life. Is there something special about this place? Or are we just not skilled enough to recognize aliens elsewhere? There are tangible answers to this question buried here on Earth, in other worlds that we share with the Sun, and perhaps even in the unbound interstellar objects that pass through our system.

We don't yet know how Earth came to be inhabited, or even whether organisms can flourish on other solar system bodies. But we are getting better at hunting for the embers of life on our neighbors. Indeed, some scientists believe we may have already glimpsed them—in spacecraft instruments, in the heart of a space rock, in the scattered light of otherworldly clouds.

The following is a state of play of the search for life in our solar system, an endeavor that is mainly focused on five targets: Mars; Venus; Europa, a moon of Jupiter; and Enceladus and Titan, moons of Saturn. These are dramatically different worlds, but that's also part of the fun. What we find on these diverse shores will tell us, in essence, just how specific an environment has to be for life to emerge—and whether there really are aliens next door.

▲ A solar family portrait. Could there be aliens lurking in our own backyard?

▶ A magnificent composite view of Marker Band Valley, a region of Mars that once hosted an ancient lake, captured by the *Curiosity* rover.

# Mars

**M**ARS IS A DESERT WORLD, AND DESERTS have a way of playing tricks on our minds. The planet's characteristic crimson glow convinced ancient peoples it must be a land scorched by fire and drenched in blood, but that was an illusion. Later generations swore they glimpsed signs of civilization on its surface, but those too were mirages that vanished under closer inspection. Mars, a planetary sphinx, has a long history of fooling us.

Now Mars is the main proving ground for the ultimate riddle: Are we alone in the universe? No other extraterrestrial landscape has attracted more public attention, resource allocation, and literal wheels on the ground in the search for alien life.

Yet in some ways, Mars is an odd choice for our astrobiological ambitions. It's only half the size of our planet, with a surface area about the same as only the land portion of Earth. Its wispy atmosphere is a hundred times thinner than the protective skies over our heads, and what little air exists is mostly made of carbon dioxide. It has no magnetic field to shield it from the onslaught of deadly radiation from space. The average temperature is a frosty -80 degrees Fahrenheit, comparable to the coldest places on Earth, and its terrain is a thousand times drier than even the most arid deserts on our planet. If life does exist on Mars, it is most likely buried deep underground, safe from the punishing conditions on the surface.

So why have we spent billions of dollars scouring this rusted wasteland? On a practical level, the answer is simple: proximity and traversability. Mars-bound spacecraft "only" have to trek across a few hundred million miles to reach their destination—an immense voyage, no doubt, but small shakes in the scope of the solar system. And while the Martian surface is desiccated, irradiated, and prone to epic dust storms, it has also proven amenable to long-term mobile exploration, unlike its foil, Venus, which has made short work of all robotic visitors that have dared to land on its surface.

There are other reasons to look for life in the Martian wilds, especially if you are not overly concerned with finding *alive* life. Indeed, the sheer deadness of modern Mars is a plus in many ways, as it has allowed the planet to immaculately preserve pristine rocks from its heyday billions of years ago, when the planet was warmer, wetter, and more welcoming to life.

Mars is visibly etched with the signatures of long-lost rivers that pooled into lakes and seas, creating aquatic environments that may have remained stable for tens of millions of years or longer. Its northern hemisphere looks like it was gradually sanded down by the ripples of an ancient ocean, creating distinctive lowlands and the specter of a vast shoreline. There are signs, everywhere, of ancient rains that once fell on this world.

It's easy to imagine that alien graveyards are sprinkled across these lands, filled with the fossilized corpses of microbes that flourished for a short time before dying out—or perhaps escaping to subterranean lairs. But while many missions have explored Mars, no slam-dunk signs of Martians have emerged yet—though they have demonstrated, beyond any doubt, that this desolate place was once habitable.

▲ A mosaic image of Mars that shows the full extent of Valles Marineris, the solar system's largest canyon, which is ten times longer and five times deeper than the Grand Canyon.

Most scientists agree that there is no conclusive evidence that life has ever existed on Mars. But "conclusive" is a high bar to clear in almost any scientific field, and it can be particularly difficult with regard to biosignatures, or biological signs of life. Indeed, over the decades, scientists have identified potential Martian biosignatures, though these clues may also have a natural geological explanation that does not involve life.

### POSSIBLE BIOSIGNATURE: *VIKING*'S "LABELED RELEASE" EXPERIMENTS

During the summer of 1976, NASA's *Viking* landers touched down on opposite sides of Mars, minting them as the operational robots on the surface of another planet. The car-sized *Viking*s were designed to search for alien life in part by scooping up Martian soil, treating it to a watery nutrient bath, and seeing if anything started metabolically kicking in the sample. The idea was that putative Martian microbes would consume this so-called chicken soup and release gas tagged with radioactive carbon, making it identifiable to the Labeled Release (LR) experiment carried by both landers.

Lo and behold, *Viking*'s LR experiments sniffed out that fateful whiff of gas. Multiple times. Moreover, when the landers heated control samples to a searing 320 degrees Fahrenheit, they did not detect the carbon gas, an outcome that matched experiments with samples from extreme environments on Earth. The implication was that the *Viking*s discovered alien life in some samples and killed it in others. (Welcome to contact with humans: You have fifty-fifty odds of being nourished or eradicated.)

For nearly half a century, the LR results have stoked debate. Was the gas produced by a chemical reaction linked to special nonliving compounds, such as perchlorates? Or was it a smoke signal from bona fide Martian organisms?

It's a notorious question mark in the annals of planetary exploration. The general consensus today is that *Viking*'s LR results are, at best, inconclusive, though some experts never gave up hope that the mission stumbled across biosignatures. Gilbert V. Levin, the engineer who designed and built the LR experiment, argued until his death in 2021 that his team had discovered the first known alien life. "I gave a talk at the National Academy of Sciences saying we detected life, and there was an uproar," said Levin. "Attendees shouted invectives at me. They were ready to throw shrimp at me from the shrimp bowl."

## POSSIBLE BIOSIGNATURE:
### ALLAN HILLS 84001

Two days after Christmas 1984, a group of scientists stumbled across a gift from Mars nestled in the Allan Hills region of Antarctica. The four-pound meteorite, known as Allan Hills 84001, seemed to have formed on Mars about four billion years ago, when the planet hosted liquid water.

Allan Hills 84001 became a media sensation in 1996 after scientists reported that some of its nanoscale features looked like "fossil remains of a past Martian biota." Many scientists have demonstrated that the observed structures could have been produced by abiotic reaction (meaning a reaction that does not involve life), prompting a debate about the ambiguous nature of the meteorite. Scientists have also reported potential signs of life in other meteorites, though none of these observations are considered conclusive evidence of aliens.

> **HOW DO ROCKS FROM MARS,** and other worlds, end up traveling through space to land on Earth? It happens when asteroids or comets slam into the surface of these planets, a process that forces bits of debris into outer space. Over time, some of this floating detritus intersects with Earth's orbit and, assuming it survives passing through the atmosphere, can fall to our world.

> "Rock 84001 speaks to us across all those billions of years and millions of miles. It speaks of the possibility of life. If this discovery is confirmed, it will surely be one of the most stunning insights into our universe that science has ever uncovered. Its implications are as far-reaching and awe-inspiring as can be imagined. Even as it promises answers to some of our oldest questions, it poses still others, even more fundamental."
>
> **—PRESIDENT BILL CLINTON**, August 7, 1996

## POSSIBLE BIOSIGNATURE:
### METHANE ON MARS

Over the course of several decades, a series of Mars missions have sniffed out traces of methane gas in the skies of the red planet. The discovery came as a surprise, as methane should theoretically deteriorate in the Martian atmosphere. The presence of the gas hinted that there must be some leaky source of the gas on the ground that was constantly replenishing atmospheric concentrations. This source could be a form of life, perhaps similar to methane-producing microbes on Earth, but it could also be linked to geological processes.

Scientists have been working for years to solve the mystery of Mars's methane using data captured in space by orbiters and on the ground by NASA's *Curiosity* rover. While these studies have revealed that Martian atmospheric methane ebbs and flows on a seasonal basis, and even over the course of a day, the mystery of its origin remains unsolved.

## JEZERO CRATER

Our first glimpse of life beyond Earth may well come from Jezero Crater, a 28-mile hole that was gouged into the Martian surface by an impacting space rock billions of years ago.

Primordial Martian rivers repeatedly spilled over into Jezero Crater, transforming it into a briny reservoir about the same size and depth as Lake Tahoe. Over time, the floodwaters carved out a rippled delta made of upstream particles that sank to the crater floor and formed clay-rich sediments with all the chemical kindling to support life as we know it.

Today, the parched bones of this delta lie at the western edge of Jezero Crater like the body of some mythical beast, complete with an unblinking eye—the half-mile-wide crater Belva—that seems to eternally gaze into space. *Perseverance* has reached this fluvial landscape and may be trundling across microbial corpses every sol (the term for the Martian day), as far as we know. We'll have to wait until Mars Sample Return returns the rover's pristine Martian packages to learn whether aliens once lived and died in this lost, muddy lake bed.

▲ The view of Jezero Crater, which contained a huge lake billions of years ago, from the underbelly of NASA's *Perseverance* rover.

## THE FUTURE SEARCH FOR MARTIANS

Past missions to Mars have revealed that the planet was once habitable, but a new generation of rovers aims to more explicitly address the question of whether it was ever, or is currently, inhabited by life. Many of the most tantalizing regions of Mars for alien hunters, including its occasionally watery polar caps, remain off-limits to exploration, due in part to concerns that missions might accidentally contaminate regions with terrestrial microbes. The effort to land humans on Mars, which is increasingly propelled by commercial space figures like SpaceX CEO Elon Musk, may eventually require a rethinking of these guidelines. If humans are able to land on Mars, we will most certainly be bringing our microbiomes with us.

### ROSALIND FRANKLIN

- **OPERATOR:** European Space Agency
- **TYPE:** Rover
- **PLANNED LAUNCH DATE:** 2028
- **ARRIVAL ON MARS:** 2029

*Rosalind Franklin*, the rover component of the broader ExoMars mission, will search for signs of life up to 6 feet under the Martian terrain. The rover will carry a drill designed to sample subterranean spaces that might be protected from the harsh radiation that bombards the surface. *Rosalind Franklin* was initially scheduled to land on Mars in 2020 but suffered a string of launch delays for a range of reasons, including the collapse of a partnership with Russia due to the nation's invasion of Ukraine in 2022.

## MARS SAMPLE RETURN

- **OPERATOR:** NASA and the European Space Agency (ESA)
- **TYPE:** Lander, ascent vehicle, return capsule
- **PLANNED LAUNCH DATE:** TBD

For decades, scientists around the world have been trying to figure out the best way to go to Mars, pick up some pristine Mars rocks, and haul them back to Earth. We are now living through the first phase of that master plan, known as Mars Sample Return (MSR), represented by NASA's *Perseverance* rover.

*Perseverance* is drilling and caching samples at its location in Jezero Crater, an ancient Martian lake bed. A future mission will land near *Perseverance* and deploy helicopters to pick up the best samples. The Mars rocks will then be loaded onto an ascent vehicle that will launch from the planet and head back to Earth. Finally, these interplanetary goods will land at a test site in Utah around 2033, assuming this daredevil sequence all goes to plan.

▲ Concept art depicting an MSR spacecraft bringing Mars rocks back to Earth.

# Womb Worlds

**EARTHLINGS ARE SURFACE** creatures. We live in a realm of sea and sky, where the Sun illuminates our days and the stars entrance us in the dark. But as we voyage deeper into space, we are learning that our exterior position, exposed to the cosmic wilds, may not be the norm in the universe.

Indeed, Earth is the only world in the solar system that wears its saltwater on its skin, like a film of sweat. Ocean worlds appear to be far more commonly located in the bellies of their worlds, hidden from view under crunchy shells of ice.

These places are officially known as interior water ocean worlds (IWOWs), but I like to call them "womb worlds" because of their amniotic qualities. Nobody yet knows if these hidden seas are habitable, let alone inhabited, but it is increasingly clear that they are abundant in the outer solar system. Womb worlds also appear to be highly diverse, manifesting in worlds with many forms and sizes, hinting at a vast range of possible marine environments.

Europa and Enceladus are the most well-known examples of womb worlds, but there is evidence that other moons of the outer solar system—including Titan, Ganymede, Callisto, Mimas, Dione, and Titania—contain liquid water underneath their crusts. Scientists think that many of our system's dwarf planets, such as Ceres or Pluto, may host these hidden marine environments as well.

Womb worlds turn many basic assumptions about alien life upside down. For instance, the search for extraterrestrials in other star systems has prioritized worlds like Earth, that host liquid water on their surfaces. That criteria limits our hunt to planets that sit in a narrow habitable zone around their stars; in our own solar system, this region roughly spans the orbits of Venus, Earth, and Mars.

*Europa*

Meanwhile, womb worlds like Europa and Enceladus are located in the far-flung outer solar system, way beyond conventional habitable zones. The discovery of these hidden oceans therefore vastly expands the number of environments where life might emerge. Scientists estimate that the sheer abundance of these worlds could potentially increase the overall habitability of the Milky Way one hundred times over, a result which confirms that we Earthlings are galactic weirdos for living on the surface.

In addition to their versatility, womb worlds may offer a level of stability and protection to life that is simply unavailable to surface biospheres. Here on Earth, we are at the mercy of asteroid impacts and solar flares and other vicissitudes of cosmic fate. But an ecosystem nestled inside a subsurface ocean would have a planetary shield from all this astronomical hubbub.

Any life-forms that exist in these subsurface seas are locked in a dark vault, blind to the great drama of stars and galaxies that is unfolding beyond their planetary walls. While life on Earth looks upward to the Sun for nourishment, organisms in these seas might seek chemical energy down in the deep seafloor.

We still don't know whether these worlds are habitable, let alone capable of supporting intelligent life, though we have plenty of room to explore this question right in our solar backyard. For now, it's interesting to consider what kind of strange inverted cosmologies might emerge from civilizations locked in these worlds—and what they might think of beings like us, that evolved under the light of stars.

Enceladus

Europa and Enceladus both have liquid water oceans flowing beneath their icy crusts.

# Venus

If Mars is a sphinx, Venus is a siren. It is the brightest object in the sky aside from the Sun and the Moon, imbuing it with an ethereal quality that has enchanted people for as long as we've looked skyward. That characteristic radiance is derived from its thick cloud cover, the stuff of heavenly dreamscapes—yet it hides a hellmouth of unimaginable proportions below.

It's always hard for me to describe conditions at the surface of Venus without laughing. It's just so aggressively hostile. The average temperature is 867 degrees Fahrenheit, hot enough to melt lead, to say nothing of the human corporeal form. Standing on the Venusian ground, you'd experience crushing pressures equivalent to those nearly a mile under the surface of the ocean. To add insult to injury, sulfuric acid falls from the skies and a day there lasts longer than a year. You almost have to admire our sister world for its radical inhospitality.

But like Mars, Venus may have enjoyed a more habitable heyday billions of years ago. After all, it is almost exactly the same size as Earth, and some models suggest it may have hosted a shallow water ocean within the first billion years of its life. Unfortunately, few traces of the Venusian deep past have survived the erosive forces that now consume the planet. Whereas the ancient history of Mars is laid bare for all to see, Venus has choked out its backstory under a sweltering atmosphere that is ninety times thicker than the skies of Earth.

Even as the clouds of Venus obscure its past, they may serve as the last remaining refuge of any microbes native to this unforgiving world. About 30 miles above the surface, temperatures level out to a

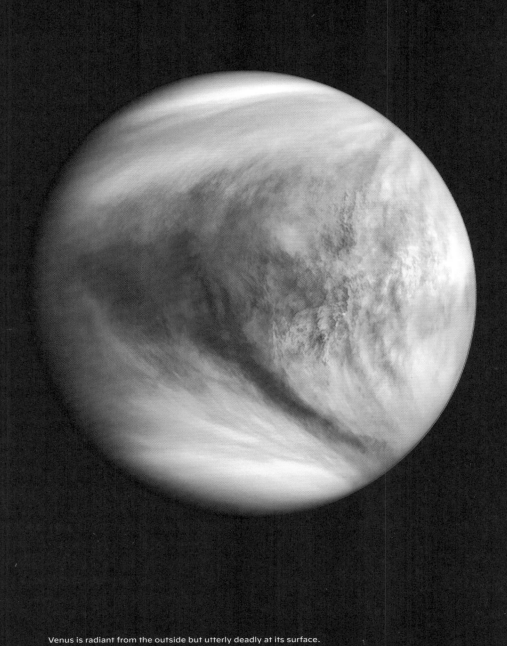

Venus is radiant from the outside but utterly deadly at its surface.

balmy 100 degrees Fahrenheit, with pressures and radiation exposure equivalent to what we experience on Earth's surface.

It's amazing to imagine alien creatures drifting through the Venusian haze, eking out a living on a rich stew of atmospheric chemicals and the faint sunshine filtering through the cloud tops. For now, it's only a dream, but some upcoming missions intend to look for skybound critters on Venus, if they exist.

## POSSIBLE BIOSIGNATURE: **PHOSPHINE**

In June 2017, a team of astronomers gazed at Venus with a telescope in Hawai'i with the aim of fine-tuning methods of detecting phosphine in the atmospheres of distant exoplanets. To their surprise, they found a detectable amount of phosphine right there in the Venusian skies, hinting at the possible "presence of life" in "a Venusian aerial biosphere," according to the bombshell 2020 study that reported the detection.

Phosphine gas is produced by some microbes on Earth as part of their digestion process, which has distinguished the compound as a sign of life. But as with methane, phosphine can also be forged by geological activity, including volcanic eruptions, so the presence of the compound on Venus's is not a slam-dunk case for life.

The discovery of Venusian phosphine inspired a wave of follow-up studies, some of which initially cast doubt on the veracity of the detection. The signal seems to have persisted in multiple studies over the years, though its source remains unknown.

▲ Pictures of Venus's scorching surface confirm that the planet is a murderous queen.

## THE FUTURE SEARCH FOR VENUSIANS

### VENUS LIFE FINDER

- **OPERATOR:** Rocket Lab/MIT
- **TYPE:** Atmospheric probe
- **LAUNCH DATE:** Late 2020s
- **ARRIVAL AT VENUS:** Late 2020s

Rocket Lab's flagship spacecraft aims to deploy a probe that will search for traces of life in the atmosphere about 30 miles above the surface. It will mark the first private mission to another planet.

### DAVINCI AND VERITAS

- **OPERATOR:** NASA
- **TYPE:** Descent probe and orbiter
- **LAUNCH DATE:** 2029
- **ARRIVAL AT VENUS:** Early 2030s

These NASA spacecraft are tasked, in part, with understanding the potential habitability of Venus in the deep past, as opposed to searching for present life in the Venusian skies.

**RELICT BIOSPHERE:** An ecological system that once flourished across a large range, but has since been restricted to a much smaller niche due to environmental pressures. Mars and Venus could potentially host underground and aerial relict biospheres, respectively, that descended from more widespread ancestors in the deep past.

# Europa

**E**UROPA IS THE SMALLEST OF THE FOUR GALILEAN moons of Jupiter. With a size just under that of our own Moon, it may host the biggest liquid ocean water in our entire solar system. Though Galileo Galilei was the first to glimpse Europa in 1610, the moon's potential habitability didn't come into full focus until NASA's *Voyager* probes snapped captivating shots of it in the 1980s. Those images revealed the smoothest surface ever seen in the solar system,

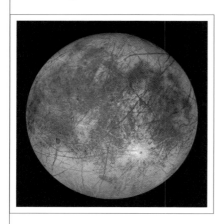

revealing a world decorated with icy cracks and few craters—all signs of a massive subsurface ocean.

Indeed, Europa may contain two or three times as much water as all the seas of Earth, making it perhaps the most promising place to look for life beyond our planet. Scientists think this ocean may be located anywhere from 6 to 19 miles under its shell of ice, and that it may extend for 60 miles, roughly ten times deeper than any point in Earth's oceans.

Jupiter's gargantuan tidal forces stretch and squeeze the moon, a mechanism that could provide warmth and energy to its seafloor. Such conditions could produce chemically rich hydrothermal vents, similar to those on Earth, at the bottom of this alien ocean.

▲ A relatively smooth and icy surface hints at the water world below.

We know that microbial communities can flourish in such environments; indeed, we may well all be the descendants of similar creatures that emerged in Earth's ancient oceans. It's also irresistible to imagine even larger life-forms, and more complex ecosystems, currently flourishing under Europa's ice shell.

## THE FUTURE SEARCH FOR EUROPANS

Europa has inspired a host of inventive mission concepts that grapple with the challenges of landing on its surface, drilling through its ice cores, and swimming in its seas. Researchers have sketched out impactors that would strike its icy surface, creating a plume of material that could be sampled by an orbiter, and proposed a parade of exotic submersibles.

▲  Europa passing in front of Jupiter's Great Red Spot. If we ever get to land on the surface of this moon, it will be quite a view.

Unfortunately it will likely be decades, or longer, before we can definitively find out if we share the solar system with space whales, or even just some alien microbes, because Europa is a hard nut to crack. For starters, Jupiter barfs out enough radiation to fry most spacecraft instruments, a feature that severely limits surface operations on its moons. Moreover, it's not clear how a probe would land on Europa's spiky terrain, let alone how it would bore through its thick ice shell and access its mysterious seas.

Of course, it's not as if we can just leave Europa hanging out there, with its alluring waterlogged tummy, and not do *anything* to explore it. That's why NASA launched the *Europa Clipper* in 2024 to gather some basic specs.

### EUROPA CLIPPER

- **OPERATOR:** NASA
- **TYPE:** Orbiter
- **LAUNCH DATE:** October 2024
- **ARRIVAL AT JUPITER:** April 2030

*Europa Clipper* will perform dozens of flybys of Europa while in orbit around Jupiter, eventually swooping to altitudes of just 16 miles above its icy surface. It is designed to search for signs of habitable conditions and to conduct reconnaissance for a possible future mission that would seek to land on the moon. The *Clipper* will also carry an instrument to sample the moon's atmospheric gases, some of which may originate from its subsurface ocean. It's possible that signs of life could be mixed into this extraterrestrial spritz.

## The Temporal Habitable Zone

**IMAGINE THAT YOU'RE** in our solar system, four billion years ago. Water flows in abundance across an ancient landscape. Sediments, rich in the ingredients for life, pile up on riverbeds and seafloors. A young Sun rises on the horizon, blanketing the terrain with its warmth and energy. We have no idea what mysterious spark creates life from inanimate dirt, but this bygone environment has all the tinder necessary to nurture and cultivate it once it does.

What's utterly wild is that this idyllic scene could apply to multiple planets in the early solar system. Earth, Mars, and Venus hosted lakes, rivers, and oceans during this long-lost era, raising the tantalizing possibility that at least three worlds in our little corner of space may have been simultaneously inhabited by life.

As the eons passed by, Mars withered into a parched husk of a planet, while Venus became a deadly hotbox in the wake of a runaway greenhouse gas effect. Earth is currently the only known inhabited world in the solar system, though as cosmic time wears on, our planet's verdancy will fade. When the Sun enters its red giant phase in some five billion years, its borders will expand hundreds of times over, swallowing Mercury, Venus, and most likely Earth.

While this will spell the end of our world, the outward advance of our dying Sun will warm the outer system, perhaps supercharging the hospitality of distant worlds that are currently beyond the solar snow line. Scientists have speculated that Titan could thaw out into a world with chilly water-ammonia oceans, or that Pluto could reach temperate temperatures and could become "the last habitable planet of the solar system."

Any life-forms that emerge on these late-stage worlds will be tragic in a way, as the youngest children of a doomed system. But if such beings were to exist in the far future, they'd be proof that life will emerge if given the chance, even around a dying Sun.

▲ A vision of Europa's ice melting away as the Sun begins its death march.

# Enceladus

ENCELADUS IS A TINY WALNUT-SHAPED MOON, measuring only about 300 miles in diameter. But this small world packs a big punch in the search for alien life. Enceladus is constantly squirting plumes of water out of an icy surface, providing ample evidence that there is an ocean swirling around under its ice shell (and also serving as the source of Saturn's so-called E ring).

The shores of these hidden seas may be only a mile or so beneath the moon's surface, making them much easier to sample than the seas of Europa. Indeed, Enceladus is practically begging for a mission to come taste its water. If microbes exist within the moon, some of them are likely to get caught up in its icy geysers. All a spacecraft would have to do is run through this extraterrestrial sprinkler and scoop up some of these accidental astronauts, if they exist.

Indeed, the now-defunct *Cassini* orbiter already road tested this idea by flying through the moon's watery gushes, though the spacecraft was not equipped to identify alien life. What's more, scientists have discovered that Enceladus contains the complete CHNOPS formula that is thought to be necessary for life. A NASA concept mission called Enceladus Life Finder aims to return to the moon with the right tools to detect any life in the spray, though the spacecraft has not yet been greenlit by the agency.

**CHNOPS:** An acronym for the six elements that are essential for life as we know it: carbon, hydrogen, nitrogen, oxygen, phosphorus, and sulfur. Pronounced like "schnapps."

▶ Saturn's moon Enceladus is quite small, but its potential habitability is mighty.

## Panspermia

**THE DISTANCES BETWEEN** planets in our solar system is incomprehensibly immense, yet rocks from other worlds still occasionally land on Earth, and pieces of our home world are no doubt embedded in many extraterrestrial vistas. This constant exchange of planetary dust and rubble raises a delicious question: Could life potentially hitchhike between worlds on rocky debris?

The concept that life could be transferred between planets, either by natural or directed processes, is known as *panspermia*, a phrase that means "seeds everywhere." Though the hypothesis has been debated for centuries, its modern incarnation suggests that extremophile microbes could survive interplanetary dispersal, a feat of strength that would involve ejection from a planet, long-term exposure to the harsh conditions of space, and atmospheric reentry and impact on another planet.

The odds of survival in such circumstances are long, but even the tiniest chance of success would have major implications for our understanding of life's origins. If microbes could travel between worlds, it would be possible that life may have initially emerged elsewhere—Mars, Venus, perhaps some unknown planet in another star system—before it inadvertently arrived on Earth.

While it's tempting to consider the idea that all Earthlings are the descendants of spacefaring alien bugs, the panspermia hypothesis is, alas, not widely supported in the scientific community. However, the specter of this idea occasionally rises again as we discover just how resilient some extremophiles can be, and of course, when strange rocks fall from the sky (see "Allan Hills 84001," page 131).

> "Who knows whether these bodies, which everywhere swarm through space, do not scatter germs of life wherever there is a new world, which has become capable of giving a dwelling place to organic bodies?"
>
> **—HERMANN VON HELMHOLTZ,**
> *a physician and physicist, in 1871*

# Titan

LAST BUT NOT LEAST IN THE HUNT FOR LOCAL life, we turn to the hazy, crazy world of Titan. Larger than the planet Mercury, Titan is also the only known moon with a substantial atmosphere, as well as the only world in the solar system that hosts stable liquid on its surface, aside from Earth.

Here's the catch: That liquid is not water. Titan is blisteringly cold, with temperatures hovering around -270 degrees Fahrenheit, well below the freezing point of water. Instead, the moon's lakes, seas, and rivers flow with fluid hydrocarbons, such as methane and ethane.

Could life exist in such an otherworldly medium? It's a long shot, but it's still worth checking out Titan just in case aliens have found a way to spin its rich prebiotic roux into a postbiotic stew. It's also possible that liquid water underneath Titan's surface might provide a more conventional staging ground for life that could be nourished by the ample organic compounds found on the ground above.

▲ Saturn's moon Titan may end up as the last vestige of habitability in our solar system before the Sun dies some five billion years from now.

## THE FUTURE SEARCH FOR TITANIANS

In 2005, the European probe *Huygens* landed on Titan, becoming the first probe to ever touch down on an outer solar system world. *Huygens* sent back a few exciting images before it gave up its ghost to the moon, leaving scientists eager to return with a bigger, better spacecraft, which is known as *Dragonfly*.

### DRAGONFLY

- **OPERATOR:** NASA
- **TYPE:** Lander/rotorcraft
- **TARGET LAUNCH DATE:** July 2027
- **PLANNED LANDING ON TITAN:** 2034

This interplanetary rotorcraft will land in a vast and mysterious region of Titan called Shangri-La, which scientists think is a sand sea rich in organic material. *Dragonfly* is expected to make dozens of flights across Titan's surface.

▲ *Huygens* snapped a few stunning pics of Titan as it descended to the surface in 2005.

# The Era of Interstellar Objects

IN OCTOBER 2017, A MYSTERIOUS OBJECT FROM another star system zoomed past Earth at about the same distance that separates our planet from Venus. 'Oumuamua, as the object is now known, was the first interstellar visitor ever spotted in our solar system—and it completely defied expectations, making it the most controversial astronomical body discovered in the twenty-first century and a lightning rod for the alien-obsessed among us.

For decades, scientists have predicted that several thousand interstellar interlopers may be passing through our solar system at any given moment. These cosmic castaways may be catapulted from their native systems by chance encounters, such as the force of celestial impacts or gravitational dances between massive bodies.

As the exiles traverse the vast interstellar void between stars, they may occasionally get entangled by the gravity of star systems. Some, like 'Oumuamua, are destined to resume their interstellar journey, while others may settle around a new adopted star, far from home.

The discovery of 'Oumuamua was hailed as a major milestone, but its strange properties quickly raised eyebrows. The object appeared to be about a quarter mile long with a reddish color and elongated proportions that have never been seen in a solar system object. It was initially traveling at about 196,000 miles per hour, clearly distinguishing it as an object that was unbound to the Sun, but then it bizarrely

sped up as it ventured into the outer solar system. Though comets undergo similar accelerations when they release gas, 'Oumuamua did not produce the telltale wispy coma that is normally associated with these so-called outgassing events.

Scientists have spent years trying to piece together a single theory that might explain all of 'Oumuamua's unusual features. Some have suggested that the object might have an exotic composition, such as nitrogen ice, hinting that it could be a chunk of a Pluto-like world in some distant star system. Another team suggested it was like an

▲ We have no idea what 'Oumuamua actually looks like, but this concept art of an elongated space cigar (or spliff, if that's your preference) has become the most popular depiction.

interstellar "dust bunny" that bounced from the coma of some far-flung comet. A pair of scientists portrayed the object as a hydrogen-rich iceberg that had been gradually conditioned not to outgas by interactions with particles in the interstellar medium.

But by far the most well-known origin story for 'Oumuamua is that it could be a piece of alien technology, such as a light sail or a spent booster, that was forged by an extraterrestrial civilization somewhere out in space. This idea has been vociferously championed by the

Harvard astronomer Avi Loeb, who first proposed it in a 2018 study coauthored by his colleague Shmuel Bialy.

Loeb's fervent belief in 'Oumuamua's alien backstory has been met with escalating dismay, and even rage, by the wider astronomical community. The consensus in the field is that the object is likely to have been natural in origin, and most scientists do not think there is enough evidence to support a technological explanation for 'Oumuamua.

But there's a bigger story unfolding behind the pique and drama: We have, indisputably, entered a new era of interstellar exploration, and we didn't even have to leave the solar system to do it.

Just two years after 'Oumuamua was discovered, an amateur astronomer named Gennadiy Borisov spotted the second known interstellar object in our solar system—and this time, its origin story was pleasantly boring. Comet Borisov, as the object is now known, is a pretty standard comet that looked and behaved like its icy counterparts in our solar system.

This leaves us with the scientific equivalent of a straight-man and an oddball—a tricky pair for drawing any broad conclusions. Fortunately, however, we are likely to discover dozens of interstellar objects in the coming years as next-generation telescopes, such as the Vera C. Rubin Observatory, come online. Scientists are also developing missions that could intercept interstellar objects that pass through our solar system, providing us with a close-up look at an entity that has crossed the stars. Perhaps one day, we might even be able to hold a piece of another star system in our hands.

In a decade, we may know how many visitors from other stars are ejected chunks of planet, or wandering comets and asteroids, or perhaps, as Loeb suspects, technological artifacts that could answer humanity's deepest questions. With that in mind, it's fitting that 'Oumuamua's name is the Hawaiian word for *scout*, as it is our first look at a mysterious population of space flotsam that will soon come into sharper focus.

## A Bug's Life (in Space)

**THAT'S ONE SMALL** step for a tardigrade . . .

In the spring of 2019, a spacecraft carrying thousands of passengers crashed on the Moon. If the astronauts had been human, they would have been presumed dead on the spot, but they happened to be tiny animals called tardigrades, or "water bears," that had been packed onto an Israeli lunar lander called *Beresheet*.

Tardigrades are famous for their ability to endure extreme temperatures, pressures, and conditions—including exposure to outer space—raising the possibility that a marooned colony of adorable Earthlings may have lived on the Moon for some fleeting period of time, as their home world rose and set over a cloudless horizon.

The *Beresheet* incident has become a cautionary tale in the field of planetary protection, a major part of space exploration that aims to prevent the transfer of life-forms between worlds. (The field of planetary defense, meanwhile, is dedicated to shielding Earth and its inhabitants from dangerous cosmic events, such as asteroid impacts.) Previous experiments have studied tardigrades and other life-forms in secure containments off Earth, but *Beresheet* demonstrated that accidents can result in the unplanned release of these organisms into alien environments.

As humans expand our presence in space, we run the risk of tarnishing extraterrestrial worlds with the earthly microbes that live in our bodies and hitch rides on our spacecraft, a scenario known as "forward contamination." There is also a chance that when we perform round trips to other solar system bodies, we might accidentally bring alien life-forms back to Earth, which is known as "backward contamination."

▲ Tardigrades have the distinction of being among the toughest and cutest animals on the planet—or on the Moon, for that matter.

||||||||||||||||||||||||||||||||||||||||||||||||||||||||||||||||

I don't think it's an accident that there's
a mirror at the heart of every large
telescope. If we want to find another Earth,
that means we want to find another us. We
think we're worth knowing. We want to be a
light in somebody else's sky.

—SARA SEAGER,
*astrophysicist and planetary scientist*

# CHAPTER SIX

# 6

# DEEP SPACE

# On the evening of October 24, 1917, atop a mountain summit overlooking Los Angeles, humanity captured its first glimpse of a world beyond our Sun.

**IT WOULD TAKE NEARLY A CENTURY FOR ANYONE TO** realize that an exoplanet had left spectral prints at Mount Wilson Observatory that night. But there, minted on a glass photographic plate at a time of global war here on Earth, was the story of a world that had it even worse: shattered, dead, torn apart by its collapsed star (also dead).

Now, more than a hundred years later, we have discovered more than 5,000 exoplanets, most of which appear to be as inhospitable as the first. There are hell planets that are hotter than the Sun. There are body-horror planets contorted into weird shapes by gravity. There are slasher planets where glass shards whirl around at seven times the speed of sound. You almost have to admire the universe's maniacal side.

Even so, there may be diamonds in the cosmic rough. The building blocks of life are sprinkled across the universe, and scientists have identified dozens of planets that are potentially habit-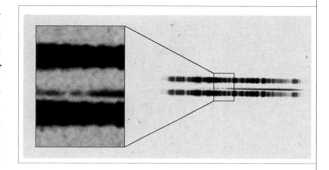able, meaning that they are likely rocky worlds that may host liquid water on their surfaces. For all we know, we have already discovered a planet that hosts life—or two, or ten. And perhaps we don't even need to directly spot an inhabited world to discover aliens: Intelligent civilizations might drop other breadcrumbs in the skies, such as beamed messages or spectral shadows cast by gigantic megastructures.

The contemporary search for life beyond the solar system encompasses a dazzling array of clever techniques to scan the cosmos for "biosignatures," or signs of life, and "technosignatures," which are signs of intelligent civilizations. Of course, none of these efforts have proved successful—yet. But this rapidly evolving field has demonstrated that the ingredients for life are ubiquitous in space, buoying hopes that our ancient quest to find *others* will one day bear some alien fruit.

> **VAN MAANEN'S STAR:** The 1917 observations recorded this white dwarf—a stellar core left over after a star has died, more or less—located just 14 light years from Earth. The object's spectrum reveals heavy elements that originated from a planet in this system. This world probably survived the death of its star, only to be ultimately shredded into pieces by the white dwarf.

▲ The first-ever evidence of a planetary system beyond our own. This glass plate shows van Maanen's star eating a planetary body and enriching the spectrum with heavy metals like calcium—a cosmic version of telltale cookie crumbs on a kid's face.

# A Note on the Absurdly Large Size of the Universe

**B**EFORE WE MOVE ANY FARTHER OUT INTO interstellar space, it's worth ruminating on the actual dimensions of our solar system, the Milky Way galaxy, and the observable universe beyond it.

The distance between Earth and the Sun, known as an astronomical unit (AU), is 93 million miles, or about 400 times the distance between Earth and the Moon.

The gravitational influence of the Sun extends out across a radius of 200,000 AU. That means the outer limits of the solar system almost reach the nearest star system, Alpha Centauri, which is 270,000 AU, or four light years, from Earth. To put that into context, if you could somehow drive to Alpha Centauri in a space-car traveling 60 miles per hour, the trip would take 7,500 years—and that's without rest stops or scenic lookouts.

We sit in a neighborhood of about 30 stars, all within 10 light years of Earth. We've already spotted exoplanets orbiting in some of these nearby systems, including Alpha Centauri and Barnard's Star. While we can't currently bridge these immense distances, it's possible to imagine a future space mission that could complete a trip on a decadal timescale.

▲ Alpha Centauri A and B, the bright stars in this picture, are both similar in size to the Sun, but Proxima Centauri is about seven times smaller, making it barely visible.

Any trips beyond this local region, however, would be an intergenerational project spanning centuries or millennia, barring unforeseen advances in space travel. The Milky Way is about 100,000 light years wide and contains an estimated 200 to 400 billion stars—and at least that many planets. Our solar system is located about halfway between the galaxy's central black hole and the outer edge of the disk. Even if we developed spacecraft that could travel at light speed, it would take tens of thousands of years to explore the far reaches of our galaxy. A spacecraft traveling at a similar speed to NASA's *Voyager* probes, meanwhile, would take nearly two billion years to cross the Milky Way, as the space-crow flies.

Zooming out from here, we get to the scale of the so-called Local Group, a gang of about thirty galaxies. Our nearest galactic neighbor,

Andromeda, is currently 2.5 million light years away. Now that we've officially entered intergalactic territories, let's remember that the speed of light is both a spatial and a temporal barrier to cosmic travel and communication. When we observe Andromeda, we are seeing light that is only just arriving after a journey through intergalactic space that lasted 2.5 million years.

In this way, looking into the sky is also a way of looking back in time. Here is a celestial tapestry filled with misleading apparitions of the past, including starlight from stars that have long since died, or lurking black holes from the early universe that may have since sprouted their own lively galaxies, somewhere out there in a "now" we can't see.

This is where it gets really trippy: Just how far can we expand our aperture? Is the universe infinite, or is there an end? Short answer: Nobody knows. But one thing is clear: The *observable* universe—meaning everything we can actually see with our instruments at this time—does have boundaries. This border stretches out across a radius of about 46 billion light years around Earth, and it's filled with hundreds of billions of galaxies.

There is a larger universe beyond this "cosmological horizon," but it will always remain out of view, because the expansion of the universe is the only thing that moves faster than the speed of light. As a consequence, there are distant stars and galaxies that are moving so rapidly, due to the expansion of space, that light from them will never be able to reach Earth.

We must now introduce the word *sextillion*, a number trailed by 21 zeros that is also known as a trilliard or a billion trillion. If you were to count every single grain of sand on Earth's beaches and deserts, it would amount to about seven sextillion grains. The number of planets in the observable universe is estimated to be about 100 sextillion at this present time, assuming that star and planet formation is similar across the cosmos.

This is simply an obscene amount of worlds, and it doesn't even account for planets that no longer exist, or future planets that have not yet been born, or planets that are beyond the cosmological horizon. Given these numbers, it can seem wildly unlikely that Earth, among this orgy of planetary abundance, is the only place that gave rise to life.

At the same time, the flamboyantly colossal scale of the universe is anathema to any conceivable method of setting out to personally explore it, or communicate with its other inhabitants, on any kind of familiar timescale. While there is some conjecture about the possibility of faster-than-light travel based on exotic phenomena, like wormholes or quantum tunneling, the current consensus is that the speed of light is, for whatever reason, a fundamental limit we must all obey.

It would be an immense challenge to launch a vessel capable of traversing these vast interstellar seas to send dispatches back from even the closest star system to Earth. There are certainly attempts in the works, but in the meantime, we need to be satisfied with becoming very skilled at looking at space from right here on Earth—and a lot of exciting news has come in on that front.

▲ A visualization of the universe's 13.8 billion-year lifespan, from the Big Bang to the vast cosmic web of galaxies we observe today. The universe continues to expand at an accelerated rate, so distant objects in space appear to be sprinting away from Earth.

▶ A deep-field view of the universe from the James Webb Space Telescope confirms that, yes, space is quite big.

# How to Spot an Exoplanet

## THE TRANSIT METHOD

**MOST OF THE** worlds identified beyond the Sun have been spotted using the transit method. This powerhouse of exoplanet discovery involves searching for very tiny dips in a star's brightness that might signal the passage (or transit) of a planet across the star from our perspective on Earth.

It's extremely challenging to see the minuscule signatures of planets amidst the blinding light of their host stars; astrophysicist Adam Frank compares it to an observer in San Francisco spying a firefly flitting around the stadium lights at Citi Field in New York City. But we now have telescopes that are up to the task, yielding a treasure trove of new worlds.

Transits reveal basic information about exoplanets, including their sizes and orbital periods, but they also offer tantalizing glimpses of otherworldly skies. Scientists hope to find potential biomarkers, such as distinct atmospheric gaseous mixes that are associated with life, in this distortion of starlight.

The GOATTs (Greatest of All Transit Telescopes) include NASA's Kepler telescope, a champion exoplanet pioneer launched in 2009 that spotted more than 2,700 new worlds using the transit method before retiring in 2018. That same year, NASA launched a successor mission, the Transiting Exoplanet Survey Satellite (TESS), that has spotted thousands more candidate exoplanets and is particularly adept at spotting Earth-scale worlds, in contrast to previous missions.

> **SPECTROSCOPY:** The analysis of a light spectrum from an exoplanet, which can provide information about its atmospheric composition, temperature, and potential signs of habitability.
>
> **LIGHT CURVE:** A measurement of the brightness of a star's light over time. Light curves obtained during planetary transits can yield information about the planet's size, atmosphere, and orbit.

## MICROLENSING

**WITHIN THE PAST** few years, astronomers have been able to detect Earth-sized planets with a scaled-down version of gravitational lensing, known as microlensing, a breakthrough that will help to reveal how common worlds like our own are in the Milky Way. Exoplanets detected through microlensing are usually positioned between Earth and a background star, causing the star's light to warp and brighten as it travels through the gravitational field of the planet. The duration of these events is linked to the mass of the lens; exoplanets usually cause stars to brighten for days or weeks, allowing researchers to make precise estimates of their masses.

## DIRECT IMAGING

**WE CAN SEE** the effect of exoplanets on their stars' motion and light, even through the perturbations of distant gravitational fields. But what about just flat-out snapping a picture of a real-life exoplanet?

While it's extremely difficult to accomplish this feat, given the blinding glare of host stars, astronomers have already managed to directly image a few especially large exoplanets, including two massive gas giants in the Beta Pictoris system. In the future, new tools like coronagraphs—which are essentially giant sunglasses made for telescopes—are expected to reveal much smaller worlds that could conceivably be home to life.

▲ Beta Pictoris b, a planet 63 light years from Earth, captured in an iconic direct image.

# What Are We Looking For?

THE MODERN SEARCH FOR ALIEN LIFE BEYOND our solar system was, at first, primarily a search for technosignatures. SETI pioneers like Frank Drake, Carl Sagan, Jill Tarter, and Nikolai Kardashev developed ambitious plans to scan the skies for signs of intelligent civilizations, including interstellar messages or hints of advanced technologies.

The SETI community endured lean years in the latter half of the twentieth century, in part due to political headwinds. SETI research was often conflated with ufology by congressional representatives, who characterized the search as a wild-goose chase unworthy of taxpayer money. Government funding for the field plummeted.

▲ The NRAO's aptly named Very Large Array stretches across more than 20 square miles in New Mexico.

Scientists then teamed up with private investors to develop facilities, which has helped propel SETI forward into the twenty-first century. Paul Allen, the late Microsoft cofounder, funded the Allen Telescope Array in California, which has been a premiere SETI research center since its completion in 2007. The Russian billionaire Yuri Milner has also invested $100 million in the Breakthrough Listen initiative, a collaboration between multiple telescopes that is the most robust effort to detect alien communications in history.

Many nations are developing their own SETI initiatives, and the field is gaining both academic and public momentum across the world. In 2016, China completed its Five-hundred-meter Aperture Spherical Telescope (FAST) telescope, which is tasked in part with searching for signals from intelligent extraterrestrials.

While no confirmed technosignatures have been detected to date, the fraction of the sky we have actually searched is "similar to the ratio of the volume of a large hot tub or small swimming pool to that of the Earth's oceans," according to one study. In other words, our failure to uncover any clear signs of intelligent life so far could have much more to do with the minuscule sample size that we've scrutinized than any broad conclusion about the rarity of civilizations.

Intelligent aliens excite our imagination, but the search is much more focused on finding biosignatures from *any* form of life. Alien civilizations, assuming they exist, are likely to be far rarer than simpler life-forms and biospheres comparable to microbes, fungi, or plants. But life doesn't have to have serious brainpower, or even a basic awareness of outer space, to leave biological traces that might be detectable to us. Some existing observatories, such as the powerful James Webb Space Telescope, are already studying the atmospheric compositions of exoplanets, and many more are slated to become operational soon.

> **RIO SCALE:** A scale developed by Iván Almár and Jill Tarter to assess the odds that an extraterrestrial signal originated from an intelligent civilization, with 0 being the most unlikely and 10 being extremely likely.

## THE CLUES WE SEEK

::: **STRANGE BREWS:** Life on Earth has profoundly affected our planet's atmosphere by altering its chemical constituents and stabilizing it over long periods. By extension, scientists hope to spot certain combinations of gases in the skies of exoplanets that might hint at the existence of life.

The simultaneous presence of oxygen ($O_2$) and methane ($CH_4$), for instance, could be interpreted as a potential biosignature. However, many geological processes produce gases similar to life-forms, so scientists anticipate challenges in unambiguously determining whether aliens are present on a planet.

One approach that could help distinguish between inhabited and uninhabited worlds is looking for signs of molecular chirality filtered through the light of exoplanet skies. A molecule is considered chiral if it is distinguishable from its own mirror image. To illustrate this concept, scientists often invoke human hands, which are near-identical but cannot be superimposed on each other (imagine trying to line up a left-handed glove on a right hand, for instance).

All life on Earth is made of these chiral compounds, meaning that our biological building blocks all have the same orientation, often described as left-handedness. Amazingly, when photons ricochet off of Earth's biological matter, these chiral molecules imprint a spiral pattern into the reflected light. Detecting these spectral spirals in the atmospheres of other worlds would be strong evidence for alien life.

::: **'TIS THE SEASONS:** Here on Earth, seasons have had a profound impact on life, and life, in turn, regulates the timing and amplitude of seasons. Seasonal changes on potentially habitable planets could expose the presence of life because they would allow scientists to search for patterns over longer timescales, which could resolve a clearer picture of biospheric processes. Indeed, you might recall (see page 130) that this is already happening in the search for life on Mars, as the red planet experiences seasonal changes in atmospheric methane that could be explained by life (and also could, unfortunately, be explained by not-life).

::: **BIOFLUORESCENCE:** There are creatures on Earth that can absorb ultraviolet light and transform it into a captivating fluorescent glow using clever little cells in their bodies. This ability, known as biofluorescence, is the reason that many animals—including birds, mammals, fish, insects, and amphibians—gleam with eerie neon shades of green, blue, and red.

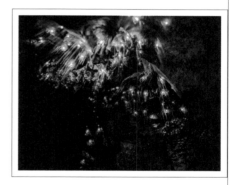

Some scientists think it might be possible to detect the spectral glow of fluorescent aliens by studying planets orbiting stars that produce ultraviolet flares, which are bursts of UV light. A world with a widespread biofluorescent ecosystem might significantly brighten as the shiny aliens eat up the energy of these flares, producing a "temporal biosignature," according to one study.

::: **COMMUNICATIONS:** The SETI field was founded on the assumption that Earth might be awash in alien messages that we haven't learned to capture or recognize. Aliens could be specifically targeting Earth with attempts at contact, or maybe we're in a good position to eavesdrop on civilizations talking to each other. Modern instruments can search for interstellar radio messages that span millions of different frequencies, while optical SETI

▲ Biofluorescent organisms absorb light at one wavelength and emit it at a different wavelength, resulting in a glow. This process is distinct from bioluminescence, which is the ability to produce light with a chemical reaction.

efforts are designed to spot high-powered artificial signals, such as laser pulses, in the optical band of light. Within the past few years, scientists have also been exploring the possibility of detecting quantum communications produced by chattering alien civilizations.

::: **POLLUTION AND PLANETARY INFRASTRUCTURE:** Industrial activities can affect the composition of a planet's atmosphere and its general appearance from afar, just as much as biological activity. In fact, you may have noticed that we are doing this right here on Earth by consuming fossil fuels that spew pollutants and greenhouse gases into the atmosphere. An alien observing Earth might be able to spot these types of technosignatures in our skies, so perhaps we'll catch a glimpse of a smoggy, sooty little world around another star. Planetwide infrastructure projects, such as city lights or thermal patterns, could also reveal the presence of an alien civilization.

::: **ADVANCED TECHNOLOGIES:** The Kardashev scale (see page 42) envisions alien civilizations so advanced that they could harness the energy of their stars, or even their galaxies. If extraterrestrial societies with these capabilities exist, we might be able to directly spot advanced technologies, including large spacecraft, mining activities, or other colossal monuments.

One common example is a Dyson sphere, a hypothetical megastructure constructed around a civilization's host star for the purpose of harvesting its radiation. This type of epic project could alter the spectrum of the star, making it potentially detectable from Earth.

Unusual light fluctuations observed in Tabby's Star, a sunlike star located about 1,470 light years from Earth, sparked speculation that the system might contain alien megastructures, such as a Dyson sphere. Follow-up observations, however, hint at a

number of natural explanations for the fluctuations, including the presence of planetary debris that is obscuring light from the star.

Researchers have dreamed up other wildly advanced technologies, such as Shkadov thrusters, that could unmoor alien civilizations from their birthplaces. These would allow them to avoid local hazards, like supernovas, while also expanding into new cosmic niches. For this reason, it may be possible to detect alien civilizations by looking for stars and other astronomical objects that don't go with the galactic flow, hinting that something else, or *someone* else, might be behind the wheel.

> **GALACTIC MOBILE HOMES:** What if alien societies became so advanced that they could move their own stars or planets to other locations in the galaxy? In the Kardashev scale, Type II civilizations are defined by their ability to just putter around the universe without actually leaving their planetary homes.

∷ ARTIFACTS: Humans have already sent multiple spacecraft on a path that will take them into the wider galaxy, to drift among the

▲ Dyson Spheres would harvest light directly from the Sun, making them both intriguing potential technosignatures and a natural plot device for a Bond villain.

stars forever or, perhaps, to be intercepted by an alien species. It stands to reason that there may be a whole slew of artifacts from other civilizations rambling through the Milky Way, like a galaxy-scale version of the Great Pacific Garbage Patch here on Earth.

The discovery of the first interstellar object, 'Oumuamua (see page 151), sparked a debate about the possibility of alien artifacts floating through the solar system. Over the next decade, a host of new instruments will be able to better spot and characterize other visiting objects that swing by our neck of the woods, while also scanning for signs of artifacts elsewhere in the galaxy.

**JUST ANYTHING KINDA WEIRD:** Scientists have some good ideas about how to look for aliens in the Milky Way based on how life and civilization have evolved here on Earth. But of course, we have no idea what extraterrestrial life might be like or what detectable traces it might leave.

For example, does all life need a genetic structure like DNA to secure its longevity across time? Does evolution by natural selection emerge on all inhabited worlds, or are there other on-ramps to complex biospheres? Can life emerge in places we don't expect, like gas giants or rogue planets? What if alien artificial intelligences are far more common in the galaxy than the biological entities that made them?

With no clear answers to these questions, scientists have to keep an open mind about the possible profiles of biosignatures and technosignatures. Of course, the universe does all kinds of weird stuff that has nothing to do with aliens. But any kind of unexplained activity on an exoplanet could help inform the search for life simply by expanding our understanding of anomalous space phenomena.

▶ A radio telescope in New Mexico collecting every bit of light it can under the arch of the Milky Way.

# Next Generation Missions

**W**E KNOW THERE ARE LOTS AND LOTS OF planets orbiting other stars, and we know how to spot and characterize the ones that are positioned just right to tip us off to their existence. Now researchers around the world are collaborating on a new generation of telescopes and instruments that can sense minute reactions in the skies of an alien world, or conversely scan huge swaths of the sky for the slightest glimmer of a technosignature.

::: **NANCY GRACE ROMAN SPACE TELESCOPE:** Set for launch in 2027, NASA's Roman telescope is designed to hunt for rocky exoplanets, some as small as Mars, using the microlensing method. Roman will also be equipped with a coronagraph that will allow it to directly image large exoplanets around planets in star systems close to our own.

::: **THE HABITABLE WORLDS OBSERVATORY:** A concept mission in development at NASA designed to directly image dozens of

▲ Astronomers expect the Roman telescope to find 100,000 new planets using the transit method and several thousand more through microlensing.

Chapter Six: DEEP SPACE ▶ 175

Earth-sized planets in the habitable zones of their stars with the help of a massive coronagraph. This telescope would be a major asset in the search for life beyond Earth, but it has not yet been greenlit and would not launch until the 2040s at the earliest.

∷ **THE EXTREMELY LARGE TELESCOPE (ELT):** As implied by its name, the Extremely Large Telescope will be a real whopper of an observatory with a primary mirror that stretches across nearly 130 feet, making it the largest eye on the sky at optical wavelengths. The telescope is currently in construction in Chile by the European Southern Observatory and is scheduled to begin observations in 2028. ELT's massive collecting area will allow it to peer into exoplanet atmospheres for signs of life.

∷ **THE SQUARE KILOMETER ARRAY (SKA):** When it is completed later this decade, the Square Kilometer Array will become the largest radio telescope ever built, with major arrays based in the Australian outback and the South African desert. The SKA is an intergovernmental effort that aims, in part, to search for technosignatures, including radio leakage from nearby exoplanets.

▲ The Extremely Large Telescope (ELT) in construction. Once completed it will be able to image exoplanets directly and characterize their atmospheres.

## The Solar Gravitational Lens

**THE SOLAR ECLIPSE** of May 29, 1919, was as much about radiance as darkness. Though skywatchers have been awestruck by eclipses since prehistory, scientists were searching for something entirely new and bizarre: Sun-warped starlight.

It was the first big test of Albert Einstein's general theory of relativity, which predicts that the gravitational fields of massive objects, such as the Sun, have the power to curve space-time. Sure enough, observations of the eclipse revealed that stars along the edge of the Sun, from our perspective on Earth, appeared distorted by the immense heft of our star. The discovery was celebrated in newspapers around the world—and helped launch Einstein into his canonical status as a cosmic visionary.

More than a century later, scientists use this trippy effect, known as gravitational lensing, as a kind of natural telescope that can zoom in on celestial objects billions of light years away. Yet its most mind-blowing application is arguably still right here at home, around the same old Sun that opened a window into space-time in 1919.

Einstein never stopped thinking about the effects of that fateful eclipse. Years later, it occurred to him that somewhere out in the deep reaches of the solar system, more than 550 times as far as Earth's orbit, the lensed edge of the Sun would converge into a focal point that could transform the star itself into a giant and powerful telescope.

The concept, now known as the Solar Gravitational Lens (SGL), has only come into focus in recent years, as next-generation telescopes have spotted more than 5,000 exoplanets. Scientists can characterize some features of these distant worlds, including their rough scales and atmospheric components, but the idea of observing them up close has remained off the table because space is, well, just very frustratingly big.

Enter the SGL. This theoretical lens would be so dazzlingly powerful that it could spy on exoplanets as if they were next door. Fine details such as continents, oceans, and, yes, signs of alien life would be clearly visible through the aperture of the Sun's gravity.

Sending a probe this far out into the solar system is a monumental challenge that demands the invention of new spaceflight technologies. For comparison, NASA's *Voyager* probes

have been traveling through space since 1977, and they are not even a third of the way to the SGL frontier.

However, the potential payoff of an SGL mission is fantastic enough to merit the effort. While the European Space Agency explored, and ultimately abandoned, this possibility in the 1990s, NASA physicist Slava Turyshev has since revived interest in a new SGL concept mission that he believes could achieve surface observations of exoplanets at astounding resolutions of just 15 miles.

It's too soon to know if such a voyage has any chance of materializing in the future. But the fact remains that this visual portal to other star systems does exist out there, far beyond the planets, just waiting for someone to peer through it.

> **THE GALACTIC HABITABLE ZONE (GHZ):** The phrase *habitable zone* is typically associated with the region around a star where liquid water could theoretically flow on a planet's surface. But in recent decades, astronomers have broadened this concept to a so-called GHZ that maps out the most likely places for life to emerge within a galaxy, like our very own Milky Way.
>
> For instance, conditions for life may not be favorable in busy clusters where stars are constantly exploding, potentially taking nascent biospheres down with them. Likewise— and this may not surprise you—living near a supermassive black hole is pretty risky, especially if it is feeding on dust and gas. These "active" black holes can blast out giant jets of relativistic particles; you wouldn't want to be on the receiving end of these cosmic death rays.

▲ Concept art of a pixelated exoplanet seen through a Solar Gravitational Lens.

# The Goldilocks Guide to Star Systems

**E**ARTH LIFE AROSE, IN PART, BECAUSE OUR Sun is a relatively stable star that is located just close enough to nourish us, but not so close that our skies boil off. There are anywhere from 200 to 400 billion stars in the Milky Way, with a wide range of masses, compositions, and lifespans. Any alien life that exists out there will be profoundly shaped by their host stars—or, in the case of free-floating worlds that roam interstellar space, the lack thereof. That's why the effort to figure out which stars might be amenable to life, and which might be hostile, is regarded as a key constraint in the search for extraterrestrial life.

- **RED DWARFS:** These stars are a giant question mark. They are by far the most common type of star in the Milky Way, accounting for about three quarters of the galactic stellar population. They are also extremely long-lived, with estimated lifespans that could stretch into the trillions of years—many orders of magnitude beyond our Sun.

  That said, red dwarfs are as small as any atom-fusing star can get, with masses that range anywhere from 5 to 50 percent that of the Sun. Because red dwarfs are so cool and dim relative

to other stars, any habitable planets in these systems must orbit extremely close to their stars, like campers huddled around feeble embers. These worlds have years that last mere hours or days. They are typically tidally locked, meaning that the same side of the planet is always facing its star (see page 183). Such tight orbits may also expose nearby planets to deadly flares and radiation, a factor that could scuttle any shot at hosting life.

The habitability of red dwarf systems is a subject of active debate among scientists, and it will take a lot more research to determine the possibility of life emerging on these worlds. But, if life *can* exist around red dwarfs, then it's reasonable to speculate that alien life is very common, given the abundance and longevity of these systems.

▲ A vision of a planet in orbit around a red dwarf, the most common type of star.

- **YELLOW DWARFS:** We know that life can exist on a planet orbiting a yellow dwarf star because you, the reader, are a life-form currently on a planet orbiting a yellow dwarf we call the Sun. The search for extraterrestrial life has been especially focused on looking for "solar twins"—meaning other yellow dwarfs, which are officially known as G-type stars—that might host planetary systems similar to our own.

    The great advantage of looking for life around yellow dwarfs is that we know at least one orbital configuration where it exists. The downside is that G-types are relatively rare, representing only about 7 percent of stars, which narrows down the possibilities.

- **ORANGE DWARFS:** Orange dwarfs, known officially as K-type stars, are often seen as the sweet spot for habitability between their red and yellow siblings. These stars are only slightly smaller than G-type stars, falling in the range of about 60 to 90 percent the mass of the Sun. They are at least three times more common than yellow dwarfs, and they live many times longer; whereas our Sun will live for about 10 billion years, K-types may shine for up to 70 billion years.

    The K-type habitable zone is located closer to these stars than in our own system, but nowhere near as close as the habitable region around red dwarfs. For this reason, planets in the habitable zone of orange dwarfs are considered particularly alluring in the search for life, because they are not likely to be tidally locked and are (theoretically) not as exposed to stellar radiation.

- **BIG STARS:** Stars that are significantly bigger than the Sun (like types B and O) can have vast habitable zones, but the emergence of life in these systems is limited by their very short lifespans. Large stars live fast and die young, usually exploding as radiant supernovas before they hit their ten millionth birthday. Any life that takes hold in these systems is probably destined to be incinerated by the death of its host star before it has any reasonable chance at escape.

::: **WHITE DWARFS:** When the Sun dies in about five billion years, it will become a collapsed husk known as a white dwarf. Some scientists have speculated that planets orbiting white dwarfs might actually host aliens that are able to survive the deaths of their stars.

▲ A Wolf-Rayet star, known as WR 124, imaged by the James Webb Space Telescope. These stars sure are pretty, but they live for only a few million years, making them unlikely hosts to aliens.

# The Goldilocks Guide to Exoplanets

THE NEW ERA OF EXOPLANET DISCOVERY HAS revealed a staggering census of planets, from giants that hug their stars to compact rocky orbs riven with diamonds to "Super-Earths" that may host expansive oceans. A small fraction of this trove of new worlds are considered potentially habitable, meaning that they might host water on their surfaces. Scientists are homing in on more detailed studies of these known worlds, while also assessing the potential habitability of other types of exoplanets that remain hidden from view. Here are some of the most interesting varieties.

∷ **EARTH TWINS:** There's no place like home, our resplendent planet Earth, but there could be a whole lot of other planets that are Earth-ish. Recent studies have estimated that one in four sun-like stars is orbited by an Earthlike planet, though not all of those worlds are situated in the habitable zone. Even so, it's

▲ Planets similar to Earth may still end up with very different environments and, potentially, biospheres.

conceivable that there are several billion planets that have comparable sizes, masses, and orbital periods to Earth in the Milky Way alone.

::: **TIDALLY LOCKED WORLDS:** Tidal locking occurs when the gravitational interactions between two massive objects end up slowing down the rotation rate of the smaller object until it matches the orbital period. For instance, the Moon is tidally locked to Earth, meaning that its rotation on its own axis takes roughly the same time as its orbit around our planet (about 27

▲ Extreme temperature gradients on tidally locked planets may create bizarre surface features resembling a gigantic eyeball. These eyeball planets are staring at you right now.

days). That's why we always see the same "near-side" of the Moon—with its big splashy features that look like a face, or a bunny, and all manner of other pareidolia—while the far side gazes perpetually out into deep space.

Tidally locked planets are common in the habitable zone of other stars, particularly red dwarfs, but the prospect of life on these worlds is still poorly understood. It's hard to imagine many organisms that could survive either perpetual sunlight or eternal darkness, but perhaps life could emerge in the twilight margin where day meets night, known as the terminator. This possibility has inspired the delightful phrase "terminator habitable" to describe these tantalizing zones on tidally locked planets.

- **SUPER-EARTHS:** One of the biggest revelations from the era of exoplanet discovery is that Super-Earths—planets that are more massive than Earth, but smaller than gas giants like Neptune or Uranus—are incredibly common in other star systems. Indeed, our own solar system is a bit of an outlier in that it consists of four rocky inner planets and four gassy outer planets, with nothing in between those extremes.

  Super-Earths are considered promising places to look for life, as models suggest that they could maintain sustainable atmospheres, support liquid water oceans, and generate the same kind of tectonic activity that has nurtured life on Earth.

- **SUPERHABITABLE WORLDS:** Earth is pretty dang habitable. Look around our planet and you will see life bursting from the seams in all kinds of environments. Life has even managed to survive in extremely weird places, such as the deepest darkest seas, scalding hot springs, or in frigid lakes miles under the Antarctic ice. But is it possible that there are planets that are even more welcoming to life, and even more densely populated by diverse creatures, than our own?

The hypothetical existence of such "superhabitable" worlds was first proposed by physicists René Heller and John Armstrong in a 2014 study in *Astrobiology*. The pair argued that the search for life in other star systems should have a more "biocentric" approach that extends beyond finding Earthlike worlds that orbit in a traditional stellar habitable zone.

"From a potpourri of habitable worlds that may exist, Earth might well turn out as one that is marginally habitable, eventually bizarre from a biocentric standpoint," Heller and Armstrong said in the study.

In other words, Earth—even with its abundance of life—could be a biological slouch compared to superhabitable planets. The team speculated that this enhanced lifey-ness (my words, not theirs) might be more likely to arise on worlds that are two or three times more massive than Earth, in part because they simply have a lot more available surface area available for alien occupation. These planets have higher odds of developing strong tectonic activity, thick atmospheres, and protective magnetic fields, which are all influential factors in habitability.

## The Unknown

**SURE, IT'S COOL** that humans have discovered thousands of exoplanets in the span of a few decades. That's a good start! Now we just have to find the other 800 billion, or whatever absurd number ultimately describes the galactic planet population. Perhaps life is most common on ice worlds, or on seafloors, or in niches that have never been imagined. Looking for Earthlike environments is a reasonable guiding principle in the search for life, but it's safe to assume that planets come in many flavors that remain unknown to us.

# The Most Exciting Worlds Discovered So Far

## PROXIMA CENTAURI B

**MASS:** **1.3 Earth masses**
**DISTANCE:** **4.24 light years from Earth**
**YEAR:** **11.2 Earth days**

**IF WE EVER** launch a mission to look for life beyond the solar system, Proxima b is the obvious destination. Located in the Alpha Centauri system, a trio of stars located about 25 trillion miles away, it is the closest potentially habitable exoplanet to Earth.

It's not yet clear whether Proxima b has the kind of amenities that are associated with life, such as an atmosphere or liquid surface water. The planet also has some potential drawbacks: Its host star is a red dwarf that spits out dangerous flares and ultraviolet radiation, though one study posits that "life could cope" on the planet, even under such extreme conditions.

Still, Proxima b's sheer proximity has distinguished it as a target for interstellar missions, such as the Breakthrough Starshot project, which aims to send a fleet of small probes to Alpha Centauri using huge laser arrays. It has also attracted scientific attention to this star system from scientists, who dream of one day mapping its surface features. The planet has even made it into science fiction films like *Invaders from Proxima B*.

▲ Concept art of the surface of Proxima b. Aliens not included.

## TRAPPIST-1 SYSTEM

MASS: **Comparable to Earth**
DISTANCE: **40 light years from Earth**
YEAR: **Anywhere from 4 to 18 days**

**THE TRAPPIST-1 SYSTEM** contains seven known exoplanets, three of which are firmly in the habitable zone of the star, making it the reigning heavyweight champion of potentially hospitable systems. While the inner- and outermost planets in the tiny system are probably hostile to life, the trio in the middle—known as TRAPPIST-1e, f, and g—are all Earth-sized rocky planets that could conceivably host water on their surfaces.

This is a very small system centered on a host star, TRAPPIST-1, that belongs to a fantastically named class of "ultracool dwarfs" because its surface temperature is lower than 4,400 degrees Fahrenheit, making it a bit chilly for a star (even though it could still incinerate you). As a consequence, the planets are all huddled close together, with years that range from 1.5 to 19 Earth days and are likely tidally locked to their stars.

This raises the tantalizing vision of a system that may contain multiple inhabited worlds, each of which looms large in the skies of its neighbors.

## KEPLER-442B

MASS: **2.3 Earth masses**
DISTANCE: **1,206 light years from Earth**
YEAR: **113 days**

**HERE IS AN** exoplanet MVP. Kepler-442b is a Super-Earth orbiting right smack dab in the middle of the habitable zone around a K-type orange dwarf that is projected to live to the ripe old age of 30 billion years (about three times longer than our own star).

This planet is located far enough away from its star to avoid getting fully tidally locked, though it probably still has a much slower rotation rate than Earth. If life has emerged on Kepler-442b, however, it will be difficult to detect or communicate with it because of its distance from Earth. It would take at least 2,400 years to conduct a basic two-way conversation.

# The Extragalactic Frontier

THE SEARCH FOR ALIEN LIFE IS ALMOST entirely focused on targets inside the Milky Way, but our galaxy is only one of billions in the universe. Some scientists are now widening the scope of SETI sky surveys to detect extragalactic technosignatures.

In 2024, for example, an international group of scientists announced the results of a massive search for civilizations in some 2,800 galaxies beyond the Milky Way. While the team found no identifiable technosignatures in their initial survey, the experiment placed constraints on future efforts to look for intelligent aliens in other galaxies.

These signals would have to be enormously loud and powerful in order to traverse the immense space between galaxies, but that might also make them particularly distinguishable as artificial. If we were to receive such a signal, there's no clear way that we could respond to it with current technologies. In addition, because other galaxies are millions or even billions of light years away, any messages that originate outside the Milky Way could come from civilizations that have long since gone extinct. On the other hand, maybe death is just an Earth thing and some form of immortality is the norm for alien societies in space. When trying to envision aliens, truly anything is possible, including entities that may outlive their own home star systems.

◄ Maybe if all the aliens in the so-called Black Eye Galaxy shouted at once, we might hear them.

||||||||||||||||||||||||||||||||||||||||||||||||||||||||||||

> Every time we tell a story about what will happen in the event of a detection, or contact, we retell the story of contact here on Earth.

—**KATHRYN DENNING,**
*anthropologist and archaeologist*

||||||||||||||||||||||||||||||||||||||||||||||||||||||||||||

# In October 1938, an actor named Frank Readick was taking notes on how to report on an alien invasion.

**HE LISTENED OVER AND OVER TO A BROADCAST FROM** the previous year describing the *Hindenburg* erupting in flames and crashing on a New Jersey airfield, killing thirty-six of its passengers and crew members. The radio reporter Herbert Morrison's live on-site coverage became iconic after the transoceanic vehicle spiraled into a catastrophic tragedy. Morrison, his voice shot through with horror at what he was witnessing, uttered the famous words "Oh the humanity!" as the microphone captured the screams and grief of the crowd gathered at the site.

The recording was Readick's inspirational grist for an upcoming episode of the weekly radio series *The Mercury Theatre on the Air*, created and hosted by the twenty-three-year-old wunderkind Orson Welles. The series dramatized classic literary works, and the October 30 broadcast would feature an adaptation of H. G. Wells's *War of the Worlds*. Readick was cast as a reporter live at the scene where the invading Martians arrived in New Jersey, of all places, before he went radio silent as one of the invasion's first casualties. He hoped to channel the surprise and panic of the *Hindenburg* broadcast as part of his performance. It's fair to say, in retrospect, that Readick and his cast members really nailed their roles.

In fact, they were a little *too* good. Welles's choice to format the drama as a fake news report convinced some listeners that hostile Martians were, in reality, marauding around the Eastern seaboard in search of fresh new human bodies to chug-and-crush like beer cans.

"The broadcast . . . disrupted households, interrupted religious services, created traffic jams, and clogged communications systems," reported the *New York Times* on Halloween, the following day. The article described "a wave of mass hysteria" that included families fleeing homes and police stations overrun with panicked calls. Local officials reassured concerned members of their communities that the broadcast, despite its careful verisimilitude to real news bulletins, was a dramatization.

It was, indeed, *all* an imaginary affair—from Wells's novel to Welles's adaptation to the actual media reports of panic.

▲ The frightening vision of contact in *The War of the Worlds* has been continually reinvented for new generations, including the 2005 film adaptation pictured here.

In the decades that followed the infamous broadcast, scholars have suggested that the accounts of hysteria and chaos were wildly overstated by the press, and that very few listeners had actually been fooled by the drama. But in a tale that sounds like it could be eerily grafted onto the modern era, the newspaper industry was experiencing devastating losses in advertising revenue to a new medium, radio, and had every incentive to launch a broadside against the credibility of its rival.

It is true that a small number of people were alarmed and critical of the fake news format. One listener filed an unsuccessful $50,000 lawsuit against CBS for causing "nervous shock," and the Federal Communications Commission instructed radio networks not to use the format of fake news bulletins again. But the more serious accounts of hospitalizations and even deaths in the fallout of the broadcast have been debunked, while claims of "mass hysteria" are largely unsubstantiated.

The broadcast was, in retrospect, one of those perfect storms when seemingly unrelated forces converge together to tear a singularity of meaning into history. For years following the misadventure, it served as a premonition of how people might respond to an actual alien attack and seemed to spotlight the vulnerability of a credulous public. Today, it is a powerful example of how media coverage can dramatically distort both our immediate perception, and long-term collective memory, of a single event.

Welles's broadcast is a fitting gateway into the fascinating question of how human civilization would respond to the discovery of alien life and what it might mean to our species to learn that we aren't alone. We've imagined so many variants of first contact moments, but it's not clear that any of them will prepare us for the real thing. In an era of media silos and rampant disinformation, any discovery of alien life will be refracted across an infinite regress of subcultural fun-house mirrors. So get ready for those post-detection grifts, because people will be hawking alien armor and cosmic protein pills within minutes of any confirmed contact.

# "First Contact" Moments on Earth

**H**UMAN HISTORY IS FULL OF EXAMPLES OF anthropological "first contact" moments between cultures that were separated by hundreds or thousands of years. Many people involved in the search for alien life look to these past encounters between divergent societies as a kind of preparatory guide for human contact with extraterrestrials.

The terminal station of this particular train of thought can be a rather drab place. After all, humans have this disturbing habit of murdering each other during such encounters. That's not to say that two cultures have never established friendly relations at first contact, or learned to maintain peaceful coexistence over time. But we have, at best, a sketchy track record filled with fraught meetings that devolved into violence and exploitation.

Indeed, it has become especially common for scholars to point to European colonialism as the ultimate cautionary tale for a post-detection world, as Indigenous peoples suffered immense tragedy in the fallout of contact with Europeans. These analogies tend to cast figures like Hernán Cortez or Captain James Cook as the "alien" invader armed with advanced technologies, while Indigenous peoples embody humanity, vulnerable to annihilation. As the renowned physicist Stephen Hawking put it, "if aliens ever visit us, I think the outcome would be much as when Christopher Columbus first landed in America, which didn't turn out very well for the Native Americans."

It's fair to assume that any aliens capable of showing up on Earth would be technologically superior to humans based solely on the capacity for interstellar travel. But would such beings be animated by the same kinds of imperial ambitions as Columbus or Cortez? Or would they spread a message of galactic connection and peace, as with human analogs like Jesus Christ or Sojourner Truth? Or would they just need Earth as a rest stop to touch some grass and do some yoga before heading back onto the galactic road, completely ignoring us?

In the absence of any information, it's easy to make arguments for any possible motivation. Perhaps any civilization that can jump across stars must exploit planetary resources on a massive scale, a situation that could put us at risk of getting plundered. Perhaps a civilization smart enough to explore the galaxy is necessarily mature enough to avoid trampling the life-forms it encounters (one hopes). Or maybe aliens might favor a certain nation or ethnicity, leading to alliances that exacerbate or overturn current international power dynamics.

▲ Contact between human cultures is often bloody, like this depiction of rolling heads and disembodied limbs at the siege of the Aztec city of Tenochtitlan.

At the end of the day, the aims of any intelligent aliens we meet are likely to be completely opaque to us, and we may never establish any substantial understanding beyond awareness of each other's existence. Kathryn Denning, an archaeologist and SETI researcher, has argued that these "direct contact historical analogies are likely to be useless" because of such ambiguities. In her view, they may actually cause harm because they can perpetuate and oversimplify real historical events that have been mangled by cultural and political biases.

In other words, we are still unearthing past perspectives that upend our traditional narratives of cultural contacts and their repercussions for the present. With that in mind, how can we possibly expect these slippery historical analogs to contextualize a discovery of alien life?

Probably we can't. But the instinct to endlessly excavate new truths and angles about ourselves, both literally from the ground and figuratively from our brains, is itself a worthy end.

## The Brookings Report

**THE US GOVERNMENT** first officially commented on the possibility of contact with aliens in 1960, at the dawn of the Space Age, with a report commissioned by NASA and the Brookings Institution. Informally known as the *Brookings Report*, the document attempts to anticipate how the discovery of extraterrestrial life might impact science, society, and culture.

"The knowledge that life existed in other parts of the universe might lead to a greater unity of men on Earth, based on the 'oneness' of man or on the age-old assumption that any stranger is threatening," the report speculated.

The authors also added, however, that societies in the past have "disintegrated" under the pressure of different ideas and worldviews.

# Possible Post-Detection Scenarios

**T**HERE IS A WHOLE FIELD OF EMERGING research devoted to gaming out the likely public reactions to various "first contact" scenarios and the best official responses to this kind of monumental event. Here are three different visions of post-detection worlds and what they might mean for humanity.

### SCENARIO 1: **FOSSILS ON MARS**

We are currently in the middle of a multibillion-dollar, decades-long plan to retrieve space rocks from some astrobiologically juicy terrain on Mars. This has never been done before, and it's a lot harder than it might sound. While a number of space missions have been able to pick up some space dirt from the Moon and a handful of asteroids, getting rocks back from Mars requires numerous spacecraft working together to pick up the goods, load them onto a craft, escape Martian gravity, and ferry them back to Earth without any risk of contamination at any stage of the journey.

◀ *Opportunity* looks back on its journey. The NASA rover established that Mars once had water and potentially habitable conditions, paving the way for future Martian fossil hunters like the *Perseverance* rover.

In other words, the Mars Sample Return (MSR) mission has its work cut out for it. But let's say it all goes off without a hitch and by 2040, we finally get our hands on some pristine Martian rocks from a lake bed that appears to have been eminently habitable more than three billion years ago. How will scientists determine whether this interplanetary delivery contains life?

The samples will be scrutinized with advanced instruments, such as scanning electron microscopes, to gain insights about their composition, structure, and petrological context. Researchers will look for patterns that resemble biological life on Earth, such as cells, complex polymers, or "bioconstructions," such as tunnels or mounds, that might have been made by life-forms.

A Martian biosignature that looks *too* similar to Earth-based life may be problematic because it could imply terrestrial contamination of the samples. The fact that our search for life on Mars essentially assumes that these organisms are long dead is actually advantageous in this way; it would be harder to mistake a fossilized Martian organism with a living microbial Earthling hitchhiker, or even a recently deceased one.

On the other hand, a biosignature that is radically different from Earth life could be easier to identify as an alien life-form. Mars samples might contain, for instance, subtle crystallized patterns or geometric structures that seem to imply some kind of assembly, but scientists would have to try to reconstruct these evolutionary pathways in order to firmly rule out an abiotic origin. This can be a challenge even on our own planet, as there are open debates about whether rocks dating back billions of years contain microbes or mere geological creations.

The upshot of a confirmed discovery of microbes on Mars, even dead ones, would present strong evidence that simple life is likely to emerge wherever and whenever the right conditions exist, even if those conditions are relatively transient. The nature of this speculative Martian life would also expand our understanding of the universe's biological possibilities.

If we find that Martian fossils are very different from the earliest analogous forms of life on Earth, it might imply that there are a variety of chemical metabolic pathways that give rise to life. If early Martians and Earthlings are similar, however, it might suggest that abiogenesis occurs through relatively predictable processes, or even that life might easily spread between neighboring planets through panspermia (see page 146). It could even suggest that life on Earth originated on Mars, or vice versa, potentially making us descendants of the aliens that we seek.

## SCENARIO 2: EXOTIC LIFE ON TITAN

Sometime in the 2030s, if all goes to plan, NASA's *Dragonfly* rotorcraft will reach Saturn's moon Titan, where it will fly from location to location searching for clues about the habitability of this strange world.

The mission's scientific payload will include an instrument called a mass spectrometer that will be housed in the spacecraft's belly. Small samples from the surface will be placed in this interior chamber and vaporized to enable precise measurements of their chemical constituents. Scientists hope that these experiments will reveal the abundance of key organic molecules on the Titanian surface, which would shed light on the kind of prebiotic environments that eventually gave rise to life on Earth.

Let's imagine that *Dragonfly*'s spectrometer discovers a curious mix of complex organic molecules that exposes what may be a slow-motion metabolic process. The mission follows up with camera observations that hint at complex structures in the Titanian soil that

▲ Concept art of NASA's *Dragonfly* craft primed to search for life on Saturn's moon Titan.

could be interpreted as potential biosignatures. One hypothesis to explain the findings is that strange life-forms use Titan's voluminous stores of liquid ethane and methane as a catalyst to produce self-replicating molecules over long timescales.

Such a discovery, even if it were somewhat ambiguous, could potentially have bigger implications in our search for extraterrestrials than the detection of fossils on Mars, where life would have relied on water. The existence of life based on an entirely different biochemical origin and structure, as would be the case on Titan, would vastly expand the number of suitable habitable environments in the universe.

"Even the most primitive kind of self-organizing organic chemistry in the lakes of Titan—an early step on the road to life—would be a momentous find," the planetary scientist Jonathan Lunine said in a 2008 lecture. "It would tell us that life began independently multiple times in our solar system."

## SCENARIO 3:
## A REPEATING OPTICAL BEACON

For decades, we have scanned the skies for artificial radio emissions from technologically advanced civilizations. Increasingly, however, scientists are also scanning the skies for much louder and brighter optical signals that could be sent into space with powerful transmitters far beyond the capacities of our own civilization.

Suppose we spot a clear and unusual signature of a loud optical burst that appears to be coming from a location in our galaxy 20,000 light years from Earth. Analysis of the event reveals that it likely originated from an artificial laser system, rather than from a natural source. At this point, several premier optical facilities are tasked with observing the source location for further pulses.

About 18 months after the initial detection, another identical pulse is observed, suggesting that there is a repeating signal at the light source. After another 18-month period, a third pulse is observed, confirming that the signals are a clockwork periodic event that is likely technological in origin.

This scenario would trigger the steps outlined in the *Declaration of Principles Concerning Activities Following the Detection of Extraterrestrial Intelligence* (see page 204). World governments would be expected to channel their scientific brainpower to inform the best methods for deciphering such a message, and then communicating these results to the global public. Any attempt to send a message back to the source would be strictly prohibited, although, somewhat worryingly, that doesn't necessarily mean a rogue state or actor might not transmit a message anyway.

A discovery of this nature would be a major breakthrough regardless of whether the intent of the message was ever decoded. The detection of an artificial beacon in space would, by itself, prove that extraterrestrial intelligences exist, a revelation that would constrain estimates about the abundance of other biospheres and civilizations.

But in this example, the enormous distance to the source of the beacon would prevent two-way communication on any kind of normal human timescale. Even if humans decided to send a reply, it would take, at minimum, 40,000 years to hear back again—assuming the civilization is even still extant. So, while this type of event would answer our most fundamental question about the universe, it probably wouldn't have much of a functional impact on our daily lives. We'd still be loners over here looking at loners over there.

> **FLARES IN THE COSMIC DARK:** In addition to searching the skies for optical beacons from aliens, some scientists have pondered how we might make ourselves visible to the universe with optically bright lasers. A 2018 study suggested that near-term emerging technologies, such as megawatt-class lasers, could be used to signal our presence to aliens up to 20,000 light years from Earth.

# The Plan (for Now)

**MORE THAN 30** years ago, members of the International Academy of Astronautics produced a document known as the *Declaration of Principles Concerning Activities Following the Detection of Extraterrestrial Intelligence*. It sketches out a nine-point game plan in the event of an alien detection:

1. International consultations should be initiated to consider the question of sending communications to extraterrestrial civilizations.

2. Consultations on whether a message should be sent, and its content, should take place within the Committee on the Peaceful Uses of Outer Space of the United Nations and within other governmental and non-governmental organizations, and should accommodate participation by qualified, interested groups that can contribute constructively to these consultations.

3. These consultations should be open to participation by all interested States and should be intended to lead to recommendations reflecting a consensus.

4. The United Nations General Assembly should consider making the decision on whether or not to send a message to extraterrestrial intelligence, and on what the content of that message should be, based on recommendations from the Committee on the Peaceful Uses of Outer Space and from governmental and non-governmental organizations.

5. If a decision is made to send a message to extraterrestrial intelligence, it should be sent on behalf of all Humankind, rather than from individual States.

6. The content of such a message should reflect a careful concern for the broad interests and well-being of Humanity, and should be made available to the public in advance of transmission.

7. As the sending of a communication to extraterrestrial intelligence could lead to an exchange of communications separated by many years, consideration should be given to a long-term institutional framework for such communications.

8. No communication to extraterrestrial intelligence should be sent by any State until appropriate international consultations have taken place. States should not cooperate with attempts to communicate with extraterrestrial intelligence that do not conform to the principles of this Declaration.

9. In their deliberations on these questions, States participating in this Declaration and United Nations bodies should draw on the expertise of scientists, scholars, and other persons with relevant knowledge.

> **THE SETI PERMANENT COMMITTEE:** A study group established by the International Academy of Astronautics (IAA) tasked with examining "all aspects of possible future contact with extraterrestrial civilizations, with special reference to international issues and activities" including "technical efforts to find evidence for extraterrestrial intelligence as well as the social consequences of such a discovery."
>
> **EXOSOCIOLOGY:** An academic field designed to anticipate the myriad consequences of a detection of alien life on human societies.

The protocols offer a basic road map for responding to an alien detection, but many experts believe they need to be updated to account for modern advances, such as the rise of social media or current geopolitical tensions. Researchers have presented fascinating quandaries about the role of the military in a post-detection world, the risk of espionage or terrorism in the fallout, and the diverse cultural impacts of such a momentous event.

# How Religions Might React to Aliens

AS WE SAW IN CHAPTER ONE, MAJOR WORLD religions have been wrestling with the possibility of alien life for many millennia. These issues have reached a new level of salience in the twenty-first century, given the vast public interest in aliens and the rapid maturation of the scientific search for them. Here's an overview of some of the religions and spiritual questions raised by a post-detection world, and how theologians and faith practitioners are planning to address them.

### ISLAM

Islam has always been relatively open to the possibility of alien life, and that mindset continues in modern interpretations of the Quran. Allah is described as a creator of life and many Muslims see no reason to limit that power to Earth. The existence of extraterrestrial life also poses interesting quagmires for Islam, as the scholar Jörg Matthias Determann points out in his book *Islam, Science Fiction, and Extraterrestrial Life*.

"Important practices seem to have been designed specifically for humans on Earth," Determann notes. For instance, how would Muslims from other planets arrange a pilgrimage to Mecca? On worlds with different days, years, and orbital configurations, how would people figure out where to direct their daily prayers? And how would Muslims observe the phases of the Moon across light years, in order to mark the beginning and the end of Ramadan?

"Long rotation periods of some planets might also make fasting from sunrise to sunset seem impossible," Determann says. Indeed, imagine planning to fast until sunset on Venus, where one rotation takes 243 days. Of course, the nightmarish conditions on Venus would kill you long before starvation, but you get the gist.

## CHRISTIANITY

For many centuries, the concept of alien life was anathema to Christian theologians for a variety of reasons, including religious tenets supporting the uniqueness of Earth and humanity. Today, however, many Christian leaders have argued that the discovery of aliens would not undermine the religion's traditional beliefs and practices, and may perhaps even bolster the miraculous nature of God.

"Just as there is a multiplicity of creatures on Earth, there can be other beings, even intelligent, created by God," said Reverend José Gabriel Funes, head of the Vatican Observatory, in 2008. "This is not in contrast with our faith because we can't put limits on God's creative freedom. To say it as St. Francis [of Assisi], if we consider some earthly creatures as 'brother' and 'sister,' why couldn't we also talk of an 'extraterrestrial brother'? He would also belong to creation."

▲ Major religions have embraced the conversation about extraterrestrial life and its implications for their beliefs and traditions, including at the Vatican Observatory.

## HINDUISM

Hinduism, with its cosmic cycles of diversity and reinvention, is in many ways a natural fit for a post-detection world that might grapple with new levels of consciousness. Hindu epics even feature flying vessels and palaces, called vimāna, that transport deities between earthly and celestial realms and present a mythological forerunner of spaceflight. But opinions may differ on how extraterrestrial beings fit into the traditional Hindu concepts of natural hierarchies, which ascends from

▲ A seventeenth-century depiction of Krishna and Rukmini as a newly wedded couple in a celestial chariot.

plants, to animals, to humans, to gods. Whether aliens are subject to karma, the culmination of a person's ethical actions in life, or samsara (the cycle of life, death, and reincarnation), could also spark debate.

"The Hindu belief in reincarnation places no limits on where in the physical universe a reincarnation event might take place," the astrophysicist David Weintraub notes in his book *Religions and Extraterrestrial Life*. This implies that living creatures on Earth could reincarnate as extraterrestrial beings in the future. It also means that some people may have been aliens in their past lives, which would actually explain a lot.

## BUDDHISM

The Buddhist belief in an impermanent and ever-evolving cosmos dovetails neatly with many of our assumptions about extraterrestrial life. In fact, the astronomer Chandra Wickramasinghe not only argued that Buddhism is open to a populated universe, but that alien variants of Buddhism might enhance the survival of civilizations on other worlds.

Wickramasinghe reasoned there is likely to be a positive correlation between "non-belligerent philosophies" and the longevity of a civilization. Belligerent philosophies, however, might fall on their own swords, literally or metaphorically.

## FOLK RELIGION

Hundreds of millions of people practice various folk religions that express a range of kaleidoscopic beliefs about the origin of life, UFOs, and the existence of aliens beyond our planet. For instance, members of the Hopi and Cree communities have speculated that Star People in their legends were actually alien visitors. Ufological folklore has also been incorporated into traditional Australian Aboriginal cosmologies, among other communities.

# Alienable Rights

> "The question is not, Can they reason? nor, Can they talk? but, Can they suffer?"
>
> —JEREMY BENTHAM, *An Introduction to the Principles of Morals and Legislation*, 1789

IN AN EPISODE OF THE COMEDY SERIES *Jon Benjamin Has a Van*, the titular character, Jon Benjamin, accidentally drives the titular vehicle through a stargate at Area 51. The van emerges on an extraterrestrial planet where it hits an alien standing in the wrong place at the wrong time. Jon then becomes enmeshed in the legal system of this foreign world, baffled by the local trial process and customs.

It's an absurdist plot played for laughs, but it also raises an important question for post-detection scenarios: If humans were to make contact with extraterrestrial life, what rights or moral values should we extend to it—or demand from it? Should extraterrestrial life get special moral consideration, even at the expense of some of our fellow Earthlings? And how much human interference with an alien life-form would be considered acceptable?

These kinds of moral and ethical questions have been at the heart of science fiction stories for decades, but they are increasingly being raised by specialists in fields like astroethics.

## THE PRIME DIRECTIVE

*"No Starfleet personnel may interfere with the normal and healthy development of alien life and culture. Starfleet personnel may not violate this Prime Directive, even to save their lives and/or their ship. This directive takes precedence over any and all others."*

The Prime Directive, aka Starfleet General Order 1, is supposedly the guiding principle for characters in the Star Trek universe. In reality, the directive has become a bit of an in-universe joke because it is casually disobeyed on so many occasions that it can seem essentially optional. But it's a good start for conversations about the extent to which humans should actually interfere with an alien species in real life.

Carl Sagan once said that if we discovered life on Mars, then we should do "nothing" with the planet because "Mars then belongs to the Martians, even if the Martians are only microbes." This sentiment is a version of the Prime Directive, informed in part by the reality that human interference with our fellow Earthlings has drastically altered ecosystems in unexpected and often catastrophic ways. If we are fortunate enough to encounter extraterrestrial life in our solar system, we will have to carefully consider the effects of our impact on it, even if our goals are well-intentioned.

For instance, imagine if we found a last gasp of microbial alien life in some subterranean pocket of Mars, but realized these hardy survivors were likely to go extinct within a few million years. Would it be ethical to try to help the Martians recover, perhaps by providing sustenance or introducing them to new ranges? Or should we never interfere with an alien species, even if our restraint may be a death sentence for them?

Likewise, what if an intelligent alien species turns up on Earth with a desperate need for refuge or resources? Our response might depend on our assessment of the value of the alien's well-being in comparison to or at the expense of our own. In such an instance, humanity would have to consider how much altruism to demonstrate and whether to demand some form of reciprocal compensation from an alien species in distress.

## IS IT EVER OKAY TO KILL AN ALIEN?

NASA's *Viking* landers, the first robots to search for life on the surface of Mars (see page 128), had an onboard experiment that was designed to detect Martians by enriching one soil sample with nutrients and exposing another to deadly heat. The experiment raises a question: Is sacrificing a small number of hypothetical alien microbes worth the breakthrough discovery of alien life?

▲ Kane (John Hurt) in *Alien* inadvisably inspecting Ovomorphs. The franchise firmly endorses the view that it's okay to kill aliens in self-defense, especially if it involves an epic exosuit battle with a Xenomorph queen.

Clearly, the answer for the Viking scientists was "yes." But this approach is, at minimum, a harsh introduction to an alien species, even one that does not appear to have sentience or consciousness. Turning the scenario on its head, would it be acceptable to humanity if an intelligent alien species incinerated a sample of a few dozen people to make sure that another luckier sample population was, in fact, alive?

There are many precedents for approaching these types of thorny ethical questions. Humans have sent soldiers into battle for thousands of years under the pretext that some number of military casualties is an acceptable trade for the security of a wider civilian population.

We have also subjected countless laboratory animals (and other life-forms) to experiments that frequently cause immense suffering and typically conclude with the test subjects being euthanized to complete autopsies. The logic in these instances is that the cruelty inflicted on a certain number of test species is ultimately redeemed by the benefits of scientific insights gleaned from this research, which may mitigate suffering and reduce death in other humans and animals.

Of course, there is far from a consensus about the validity of this reasoning; millions of people reject the moral compromises that humans have made with regard to war or animal rights. This disharmony of views would inevitably color our decisions in the wake of detecting alien life. For instance, if we send missions to search for extant extraterrestrial life within an ice moon like Europa or Enceladus, would our observations and instruments have the potential—or, as with *Viking*, the explicit intent—to harm or kill some of it? And if so, are we, as a species, comfortable with the loss of some alien life in the process of discovery?

These are not easy questions, and they should be adjudicated by an audience that is much broader than the science teams actually involved in these missions. Thinkers across a range of disciplines continue to ponder these thorny issues of interspecies ethics. According to philosopher Erik Persson, "If we encounter sentient extraterrestrials on a planet we want to explore or exploit, we have to find a way to account for their interests in our decisions. Depending on the moral theory to which we adhere, we may not have to totally abandon our plans."

There is one instance, however, in which killing aliens is likely to be the only acceptable option, and that is if they seem intent on killing us first. This scenario (see page 192) is one of our absolute favorite fantasies, and we've imagined all kinds of interplanetary wars waged between aliens and humans.

In the context of an explicit alien attack, we seem to value humans—and Earth life more broadly—more than any extraterrestrial species. But are there instances in which we would consider an alien species more valuable than an Earthling?

## THE ALIEN—EARTHLING EXCHANGE RATE

Missions that are destined to land on Mars must first be subjected to an intense routine of sterilization before they depart our planet. Extremophile organisms, especially bacteria, are so hardy that it's conceivable they could hitchhike on an interplanetary voyage on the surface of a spacecraft and then hop off on alien turf, muddying up any results and potentially interfering with native Martian life.

Future missions to other potentially habitable worlds—such as Europa, Enceladus, and Titan—will also go through this stringent sterilization process to avoid, as much as humanly possible, the contamination of extraterrestrial environments with hardy creatures from home. Of course, our track record on this score is not perfect. Just ask those *Beresheet* tardigrades (see page 153); oh, wait, you can't. They already spilled onto the Moon.

There is a simple moral belief behind this ritual act of spacecraft sterilization: Alien life is more intrinsically valuable to us than the Earthling extremophiles that might adulterate it. In part, we bestow this value on alien life for its sheer novelty; the weirdest creatures we find on Earth—and boy, do we ever find weird ones—are still born from the same planet. A biological entity that emerged on a different world would offer incredible scientific insights that no species on Earth could match.

▲ Yingrui "Zao" Huang maintains the "clean room" at NASA's Goddard lab. Scientists use these contamination-free labs to assemble spacecraft, store extraterrestrial rock samples, and, yes, potentially help keep aliens alive.

|||||||||||||||||||||||||||||||||||||||||||||||||||||||||||||||||||

For a moment, nothing happened.
Then, after a second or so,
nothing continued to happen.

—DOUGLAS ADAMS,
*The Hitchhiker's Guide to the Galaxy*

|||||||||||||||||||||||||||||||||||||||||||||||||||||||||||||||||||

# Here we are, waiting with bated breath for a momentous discovery.

**IT COULD BE FOSSILS ON MARS, MARINE ECOSYSTEMS** inside Europa, an artificial message from deep space, or a chemical wink in the skies of an exoplanet. Or—who knows?—a spaceship that lands in a field one day and just disgorges a bunch of aliens.

But imagine if detection of extraterrestrials eludes us for the next decade. We keep looking. Our skills and technologies improve. A century rolls by, then two, then a millennium. It's hard to know what our civilization will look like in the year 3000. Our descendants might still be looking, refusing to lose hope. Thousands more years roll by. Humans either survive, or they don't. But whether we fizzle out in the short term or set out into the galaxy, what if the grand moment never arrives? What if none of the lines we cast off into space ever attract a bite? No microbes. No civilizations. No contact.

And then the Sun dies and takes Earth down with it.

Alien optimists have abounded for centuries, but there has always been a vocal minority of alien pessimists who have suspected either that we are literally alone or that complex life is so rare that we may as well be it. Many generations have entertained the possibility that we Earthlings occupy a special place in time, space, and cosmology. These skeptics don't see the Fermi "paradox" as paradoxical at all, but rather the logical outcome of a universe that's virtually devoid of life or simply not amenable to contact or detection.

Still, the possibility of a "no contact" future seems, to me at least, oddly melancholic, not to mention flat-out puzzling. Our galaxy has churned out billions of worlds over its epic lifetime—and the Milky Way is, of course, just one of billions of galaxies—so it seems incomprehensible that Earth is the only place (or one of the only places) where dirt somehow figured out how to live. And for all of our terrifying tales about hostile alien invaders, is there anything more bone-chilling than the thought that we really might be all by ourselves out here, a biological fluke in a sterile cosmos?

This outcome might seem, well, alienating, but it would also emphasize the extraordinary singularity of our planet as the only living world among a sea of planetary dead zones. The mystery of whether life exists elsewhere in the cosmos would be replaced by the mystery of its apparently unique emergence on Earth, and our gaze might shift from the dark expanse of space to the ground beneath our feet.

There are obvious lessons in this lonely future, including one we seem keen to avoid—we really, *really* need Earth to remain habitable—but it's also hard to predict how we might react to our search for life ending with such an anticlimactic whimper.

▲ The Arecibo Observatory was a pioneering instrument in the search for alien life and operated for more than 50 years before it was destroyed by a cable collapse in 2020.

# A Paradox Solved?

THERE IS ONE SOLUTION TO THE FERMI PARadox that haunts the peripheries of our search for aliens. It's not that we haven't looked hard enough. It's not that our messages are incomprehensible. It's not that we are being watched without our knowledge.

It's that they are simply not there.

Nothing ever quickened in the soils of the worlds that share our Sun. Nobody is there to pick up the calls we're sending into space. We are, after everything, alone.

## THE RARE EARTH HYPOTHESIS

Perhaps the most forceful modern case for this position was put forward by the paleontologist Peter Ward and the astrobiologist Donald E. Brownlee in their 2000 book, *Rare Earth: Why Complex Life Is Uncommon in the Universe*. Ward and Brownlee pull together research from multiple fields to argue that complex life is extremely unusual beyond Earth, though they suggest that simpler microbial life-forms are probably common throughout the universe.

The scientists base their hypothesis on the vastly improbable sequence of coincidences that have shaped Earth into such a lively biospheric dynamo over its 4.5-billion-year lifespan. It is at the right distance from the Sun to avoid being alternately scorched or frozen. It is big enough to sport both a magnetic field, which wards off harmful

▲ The Moon photobombing Earth, captured by NASA's EPIC camera. Our little orbiting friend stabilizes our climate, which may have helped jump-start biology.

radiation from space, and plate tectonics, which produce active geological environments and stabilize the climate. It just happened to get smacked in the face by another planet early in its lifetime, a collision that produced the Moon; this companion has enhanced Earth's habitability by stabilizing our planet's spin axis and producing tidal zones.

Earth also wasn't flung into interstellar space or the crucible of the Sun by the migrations of gas giants like Jupiter, a fate suffered by hundreds of millions of exiled or engulfed planets in our galaxy alone. Our solar system may be located in a galactic sweet spot, far from the dangers near the central black hole, but not so far out that it is devoid of heavy elements. This world has been pelted by space rocks, but its biosphere has miraculously never been totally wiped out. It is also enriched with enough carbon to spark life, but not so much to turn it into a runaway greenhouse hothouse like Venus.

The list of cosmic coincidences is essentially a winning lottery ticket. While there are many worlds that share a few of the same numbers, very few (perhaps none) have experienced exactly the right setup to make the leap from mere microbes to the complex creatures that Charles Darwin described as "endless forms most beautiful and most wonderful."

Ward and Brownlee's iconoclastic view has inspired plenty of discussion and critiques about its scientific merits, especially as new discoveries have been made over the past 25 years since the publication of *Rare Earth*. But regardless of whether they are right or wrong about the cosmic rarity of complex life, the pair offered a lesson about our stewardship of the planet.

"What if the Earth, with its cargo of advanced animals, is virtually unique in this quadrant of the galaxy—the most diverse planet, say, in the nearest 10,000 light years?" they write. "What if it is utterly unique: the only planet with animals in this galaxy or even in the visible Universe, a bastion of animals amid a sea of microbe-infested worlds? If that is the case, how much greater the loss the Universe sustains for each species of animal or plant driven to extinction through the careless stewardship of *Homo sapiens*? Welcome aboard."

▲ Earth got smacked with a Mars-size planet and still managed to become a productive vessel for life.

## THE GREAT FILTER

Another popular solution to the Fermi paradox proposes that life must clear at least one cosmic hurdle before it can mature into intelligent societies capable of spreading beyond their native star systems. The existence of this so-called Great Filter would explain why there are, so far, no obvious signs that Earth has been visited or messaged by intelligent aliens.

The concept of a Great Filter was first introduced by economist Robin Hanson in 1996 to explain the apparent absence of detectable alien life beyond Earth. Hanson suggested that the evolutionary path toward long-lived intelligent civilizations could be stymied by several high bars, such as the starting point of a hospitable planet, or the need for adaptive innovations like sexual reproduction or tool-using, or the ability to handle self-destructive technologies. Any one of these limits, or a combination of them, could filter out extraterrestrial beings on a trajectory toward a civilization with a capacity to attempt interstellar communications.

It's possible that the filter may sit at the actual origin of life, which could be cosmically rare if early forms of life, including the self-replicating molecule RNA, require very specific conditions to assemble that were uniquely available on Earth. Perhaps the filter sits at the transition from single-celled organisms to more complex forms of life, which occurred on Earth around 2.7 billion years ago and involved what might have been a vanishingly unlikely union between two simpler forms of life. In these instances, the filter lies in our past, meaning that humanity and all other extant Earthlings have already successfully passed through the bottleneck. But the Great Filter could also be situated alongside the development of futuristic and potentially self-destructive technologies; in this case, our civilization may be in the final approach toward this deadly endgame.

Many pioneers in the search for extraterrestrial intelligence were also involved in the development of nuclear weapons, including Enrico

Fermi himself. As a consequence, a fear of catastrophic self-destruction is imprinted into the DNA of the entire field. These anxieties have unfortunately been validated by the recent emergence of yet another self-induced threat, anthropogenic climate change, driven by humanity's consumption of fossil fuels.

Within the span of a century, our civilization has produced multiple dangers to itself, raising the odds that humans might wipe ourselves out before we have a chance to make contact with another species. If self-annihilation is one possible outcome for our own species, might it be among other civilizations, too? Across the universe, advanced societies may rapidly arise and just as quickly become victims of their own destructive creations.

"No alien civilizations have substantially colonized our solar system or systems nearby," said Robin Hanson in his 1998 essay on a possible solution to the Fermi paradox. "Thus among the billion trillion stars in our past universe, none has reached the level of technology and growth that we may soon reach. This one data point implies that a Great Filter stands between ordinary dead matter and advanced exploding lasting life. And the big question is: How far along this filter are we?"

As with any solution to the Fermi paradox, the Great Filter is built on speculative assumptions. For example, just because human societies have a long track record of self-destruction doesn't mean alien civilizations will face the same foibles. Maybe some planets are inhabited by intelligences that can easily resist the urge for overconsumption; these extraterrestrials might shake their heads—or whatever appendages they have—at Earthlings' inability to sustainably manage our resources and technologies. It's also possible that the human compulsion to explore and colonize new frontiers is an outlier; perhaps intelligent aliens have never arrived at Earth because they are perfectly happy with their home worlds.

In the coming years, we may find some interesting new constraints for the Great Filter hypothesis—if we find microbes, or even more complex life, here in our solar system. The philosopher Nick Bostrom has argued that the discovery of even simple life on Mars would actually be "a bad omen for the future of the human race" because it would imply the filter does not lie in the past, but in our future. "Such a discovery would be a crushing blow," Bostrom writes. "It would be by far the worst news ever printed on a newspaper cover."

## THE DARK FOREST

Humans have debated the wisdom of actively trying to contact intelligent aliens for several decades (see page 46). But what if the extraterrestrials that hear our missives turn out to be the hostile monsters that we've frequently conjured up in our nightmares? By calling out to the cosmos, are we writing our own death sentence?

Perhaps other civilizations embroiled in similar debates among their communities ultimately come to the conclusion that it is better to be silent than supper. This solution to the Fermi paradox is commonly known as the Dark Forest hypothesis, after Liu Cixin's 2008 science fiction novel of the same name—though the concept of voluntarily quiet civilizations predates the book's publication. The novel suggests that the Hobbesian vision of a natural world "red in tooth and claw" extends to an alien-eat-alien cosmos that favors civilizations smart enough to fly under the interstellar radar.

"The universe is a dark forest," Liu's protagonist Luoji explains, where "hell is other people. An eternal threat that any life that exposes its own existence will be swiftly wiped out. This is the picture of cosmic civilization. It's the explanation for the Fermi Paradox."

Under this solution, the reason why we've never been contacted or visited by intelligent extraterrestrials is simple (and perhaps justified) paranoia.

## BERSERKER HYPOTHESIS

Humans are essentially meatbags that are laughably ill-adapted for the perilous conditions of outer space. We have sent astronauts into orbit and landed them on the lunar surface, and we dream of transporting our fleshy Earthling bodies to even more distant vistas. But there's a reason that our robotic explorers have far outpaced us in the exploration of space: They don't have to eat, breathe, or do any of the cumbersome tasks associated with life.

It makes sense, then, to conclude that most intelligent civilizations, including ours, opt to send artificial scouts into space long before they ever haul their actual biological asses (or the equivalent) out of their star systems. A civilization hungry for resources might develop a whole fleet of self-replicating intelligent "berserker" robots that can travel through the galaxy, harvesting valuable

▲ Alien beserkers at work: A pessimist vision of first contact.

minerals, collecting data, and potentially killing off other forms of life.

This thought experiment has given rise to the Berserker hypothesis, which suggests that we haven't encountered intelligent life because it has been mostly destroyed by hypothetical robot killers dispatched by one, or more, alien civilizations. So why haven't we been wiped out by these terrible robots? It could be that we are simply next on the list, as the systematic extermination of all life in a galaxy is no doubt a long-term project. While this is a scary prospect, some researchers have pondered whether such a diabolical plan would be worth the effort for any E.T.

"Such a cruel action could be accomplished only by a very evil species of ETs, inspired by an almost satanic will, to the point that they desire the destruction of another civilization as an end in itself," wrote the interdisciplinary scholar Paulo Musso in a 2012 study.

## The Hart-Tipler Conjecture

**BERSERKERS ARE A** killer subclass of a larger class of hypothetical self-replicating spacecraft first envisioned by the mathematician John von Neumann. These so-called "von Neumann probes" could be deployed by advanced civilizations for many reasons beyond interstellar domination, such as reconnaissance or ambassadorial outreach.

Some scientists have gone so far as to argue that the apparent absence of any easily detectable alien probes is itself evidence that extraterrestrial intelligence is rare, or perhaps nonexistent. This concept is commonly known as the Hart-Tipler Conjecture, after physicists Michael Hart and Frank Tipler, who separately developed it in the 1970s and 1980s.

If extraterrestrials had the technology for interstellar communication, Tipler argued in 1980, "they would also have developed interstellar travel and would already be present in our solar system."

## THE "EARTH ZOO"

We typically think of *Star Trek*'s Prime Directive (see page 211) as a guideline for human interactions with other beings. But what if we are actually on the receiving end of a similar directive? What if alien societies are aware of Earth, and each other, but have remained deliberately out of sight and contact so as not to disrupt our natural development?

This concept, known as the Earth zoo hypothesis, imagines our planet and its inhabitants as a natural preserve that is under surreptitious observation by alien "zookeeper" civilizations. Many iterations of the hypothesis suggest that these observer aliens will not make contact with other civilizations until they exhibit certain prosocial characteristics or pass some technological threshold.

"Any attempt to assess [the zoo hypothesis's] worth must begin with its most basic assumption—that ETIs share a uniformity of motive in shielding Earth from extraterrestrial contact," wrote astrophysicist Duncan Forgan in a 2011 study. He therefore deems it a low probability event.

But low probability isn't no probability. Imagine if Earth really is such a backwater that aliens have banded together to make sure that nobody interferes with us. On the one hand, that's pretty patronizing, but I'll take it over the Dark Forest.

> "Sometimes I think the surest sign that intelligent life exists elsewhere in the universe is that none of it has tried to contact us."
>
> —BILL WATTERSON, *Calvin and Hobbes*

◀ The zoo hypothesis suggests that aliens are voyeurs as much as they are voyagers.

## INVISIBLE TECHNOLOGIES

Ever since Kardashev imagined galaxy-powered alien empires, scientists have been looking for extremely bright signals or massive megastructures that might expose an advanced civilization. But what if we're wrong to assume that bigger is civilizationally better in the cosmos? Perhaps aliens rely on technologies that produce no detectable emissions or structures, like quantum computing or miniaturized energy grids.

"It could be that it's the engineering of the small, rather than the large, that is inevitable," SETI scientist Seth Shostak has speculated. In other words, advanced aliens might rely on nanobots and other miniaturized technologies to meet their civilizational demands, as opposed to developing megastuctures to shuttle their stars around or create interstellar diasporas.

## THE FIRSTBORN

Many solutions to the Fermi paradox invoke the vast astronomical distances that separate Earth from other worlds, some of which may be inhabited. But what if the big limiting factor is not space, but time? What if Earth is a biological early bloomer in a universe that will experience its most fertile years billions of years in the future?

In this way, the Firstborn hypothesis imagines Earthlings as the elder children of a cosmos that has not yet entered its most productive biological phase. This possibility lines up nicely with the view, put forward by some scientists, that the universe is getting more habitable over time. Over the course of its 13.8-billion-year lifespan, the universe has generally calmed down and stabilized, producing far fewer galactic death rays while also successively enriching itself with the ingredients for life with each new generation of star.

Some scientists have also pointed out that red dwarfs, those small long-lived stars, could nurture life on their planets for trillions of years, hinting that more complex aliens may take untold eons to spring up in these systems.

In short: We turned up way too early to a cosmic party that will likely start raging long after the extinction of our own species. Putting aside the cosmic FOMO of this scenario, it's intriguing to consider how this position might, in time, change humanity's perspective about its role and relationship to the wider universe.

We have spent decades searching for elder intelligences in space, but what if we are the firstborn we seek? Do we then have a responsibility to try to communicate with other species across time, in addition to space? Will sentient aliens in the far future decode the secrets of our civilization through our artifacts, as we do with our own lost ancestors? What will they make of that Doritos ad we sent into space (see page 50)?

We look for alien life in the stars because we want to learn from other civilizations, but we may be most valuable as teachers. Even if these future life-forms are advanced relative to humans, they might welcome information about our perspective on the universe. Just imagine what a boon it would be for Earthlings to receive this kind of message-in-a-bottle from a bygone alien society, regardless of whether or not it was technologically superior to us.

Firstborn children are often tasked (fairly or not) with setting an example, offering guidance, and protecting their younger siblings. Humans may have to confront our lot as the first cosmic example of "oldest child syndrome."

||||||||||||||||||||||||||||||||||||||||||||||||||||||||||||||

You may forget but
let me tell you
this: someone in
some future time
will think of us

—SAPPHO, *ancient Greek poet*

||||||||||||||||||||||||||||||||||||||||||||||||||||||||||||||

# CHAPTER NINE

# WHEN WE BECOME THEM

# In about 40,000 years, the inhabitants of a planet orbiting a dim red dwarf star will discover the subtle glint of an alien artifact passing close to their system.

**A DECISION IS MADE TO DISPATCH A PROBE TO COLLECT** the foreign object and bring it back for closer study.

The mission is a success and the artifact becomes a global sensation. Whole new fields of study are pioneered for the purpose of understanding its instrumental components and, especially, the mysterious circular ornament, inlaid with a valuable star-forged element, that it carried from its unknown home. What was its purpose? What were its makers trying to say?

This is a possible future that may await *Voyager 1*, humanity's first interstellar probe, when it reaches the humble star Gliese 445. Nobody knows if there are planets around this red dwarf, let alone if they host life. But *Voyager* is on track to encounter the system at a point as far into our future as the presence of our Neanderthal kin is in our past. This artifact will represent human civilization in the 1970s, a snapshot in time of our technology, our culture, and our desire to connect with life-forms beyond our world.

It takes two to make first contact. A dazzling infinity of possible alien encounters awaits us in the future, but it will always be us on the other end of the line. We should continue to daydream about what it will be like when we find them, but it's also worth asking what they might find in us.

We have, after all, sent multiple messages into space, both as physical entities and radio light, that will traverse the galaxy for thousands of years to come.

In the future, we may even find new homes for ourselves among the stars by spreading human civilization, and other forms of Earth life, across interstellar space. Just as we have yearned to find alien life, so too have we long dreamed of leaving our planet behind to explore, and settle, new worlds. Though we have taken a few steps into this great expanse, it remains to be seen whether we will ever be able to support a population beyond Earth. Still, as a species that has landed on the Moon and inhabited space stations, we are the first example of the extraterrestrial explorers that we seek.

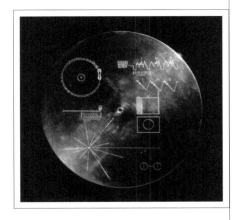

We can imagine many possible human futures, including a multi-planet civilization, an epic odyssey on a generation ship, or an Earthbound society destined to extinction. What would it be like to cast humans in the role of interstellar voyagers that show up, unannounced, in the skies of some distant inhabited planet? And what would be left of our civilization on Earth for future life, either Earthling or alien, to find?

▲ The *Voyager* records may become the most long-lasting creation of our civilizaton.

# Homo spaciens

**M**ORE THAN 700 PEOPLE HAVE VISITED OUTER space since Yuri Gagarin first blazed a trail to this new frontier in 1961. While twelve of these astronauts reached the surface of the actual Moon, the vast majority of space travelers have ventured no farther than 250 miles above the planet's surface, in low-Earth orbit.

We've learned a lot from these baby steps into the great beyond, but the effort has cost hundreds of billions of dollars and, more crucially, the lives of nineteen brave explorers. And while we have made progress in understanding the treacherous environment in space, it's clear that the Space Age vision of a thriving civilian society beyond Earth will require such staggering expenses and sacrifices that it may never become a feasible reality at all.

Yet if we humans want to escape extinction, we will eventually have to set out for other worlds. Earth is a temporary haven. It's only a matter of time before it is struck by another space rock that is as devastating—perhaps more devastating—than the one that killed the dinosaurs. Even if we learn to defend our planet from such impacts, while also reigning in our own destructive behaviors, the Sun will eventually enter old age, brightening and expanding until it imposes a sterilizing quiet onto this lush world.

There are two main ways that we might escape the perils of Earthbound life: We can either adapt our bodies to space, or we can adapt space to our bodies. In the first instance, known as pantropy, we might ditch our fleshy bodily vessels in favor of a new artificial form

Chapter Nine: WHEN WE BECOME THEM ▶ 237

better suited for the harsh world beyond our planetary cradle. Science fiction is filled with examples of biological entities that transfer their consciousness to robotic structures, though it remains to be seen if such technologies will ever materialize in real life.

The second prospect is generally considered more palatable, as most humans are understandably attached to their familiar bodily form. In this version of a post-Earth future, humans might build cities in space stations, or terraform planets, or send generation ships to the stars. Humans would then continue to evolve to fit new niches, becoming what the author Frank White calls a civilization of *Homo spaciens*.

▲ Astronaut Michael Gernhardt (foreground) pictured with Earth (everything else) on Space Shuttle *Endeavour* in 1995. God bless robot arms in space.

▶ The International Space Station. Isn't it wild that people have been living in that thing for more than 20 years straight? What a goofy species.

## OFF-EARTH COLONIES

One of the keys to humanity's remarkable rise is our ability to adapt environments to us by building shelters to protect from otherwise hostile conditions. From ice-bricked igloos to alpine temples to the many vessels that carried us across seas, we have shown an aptitude for encasing ourselves in artificial bubbles of habitability to explore new horizons.

Space is the most challenging frontier we have ever faced, but we've learned some off-Earth fundamentals with the help of our crewed research outposts, especially the International Space Station. Creating a space station that could house large numbers of civilians, however, is far beyond our current capacities. In the meantime, there are many compelling concepts on the drawing board.

∷ **O'NEILL CYLINDERS:** Humans have evolved to live on Earth, and any colony we hope to inhabit with our regular bodies will have to simulate an Earthlike environment. The O'Neill cylinder, and its variants, offer a popular concept to make that leap.

▲ A vision of the inside of an O'Neill cylinder, where artificial gravity would be modulated by the rotation of giant toruses.

Named for the space visionary Gerard K. O'Neill, who first popularized them in the 1970s, these stations would be shaped like a huge space donut, allowing them to produce artificial gravity with centripetal force by rotating at a controlled speed. Habitable environments—including farms, cities, and bodies of water—could be constructed in the bottommost surfaces of the stations, supporting human populations that could either stay in the solar system or venture to other stars.

- **GENERATION SHIPS:** If humans ultimately decide to sail the immense interstellar seas, they will need to be prepared for a very long journey (barring the discovery of faster-than-light travel). For this reason, some space visionaries have proposed the development of spacecraft capable of sustaining human crews for voyages that may last tens of thousands of years.

  These "generation ships" are so named because countless generations of humans would be confined to them before they reached the shores of any alien exoplanet. The initial crew that left Earth might seem like fabled gods to their descendants, as the human life cycle repeats in its staid borders out there in interstellar space. The offspring of each new generation would be the ultimate "indoor kids," having no opportunity to bask in the glow of a Sun left behind long ago, but they could also be the progenitors of a human civilization that spans light years.

- **TERRAFORMED WORLDS:** Our planetary neighbors, including Mars and Venus, are fixer-uppers, to say the least. This reality has raised the possibility of terraforming, a hypothetical process of engineering other worlds to suit our human needs. Terraforming advocates have suggested everything from detonating nuclear weapons on Mars, with the intent of kicking up an atmospheric blanket, to seeding the clouds of Venus with

photosynthetic bacteria that could deliver breathable oxygen to this world.

Terraforming projects would be the most massive infrastructure works ever conducted by humans. If successful, they could allow our species to expand to occupy many new niches, from Martian caves to Venusian skies to the icefields of Europa.

**PLANETARY CHAUVINISM:** The fixation on humans as a species that must inhabit planets, rather than space stations or other environments.

▲ Concept art of Mars being terraformed over time. Dibs on the oceanfront property.

# Do We Stay or Do We Go?

**H**UMANS HAVE RADICALLY ALTERED THE TRAjectory of life on Earth. Our technologically advanced society has pushed thousands of species into extinction and made hundreds of others dependent on our communities. As we envision a human future beyond our planet, it's important to ask the question: How much should we alter alien worlds we encounter? And given the apparent rarity of life beyond our planet, do we have a responsibility to spread living beings beyond it, even in places where they might never naturally emerge?

The ethics of space exploration have become a hot topic in recent years as NASA plans to return humans to the Moon as part of its Artemis program. Meanwhile, private space companies are actively eyeing destinations like Venus and Mars. Spacefaring entities are subject to a handful of treaties that encourage the peaceful and sustainable use of off-Earth environments, but these loose guidelines may not hold if major space players—be they governments or corporations—start testing them.

The Moon's surface already contains multiple archaeological sites associated with the Apollo Moon landings, as well as other robotic visitors to its surface. Future missions may attempt to proceed with large-scale resource extraction or even human occupation, which would dramatically change the lunar environment.

There are diverse views about the extent to which humans should fashion alien environments to fit our needs. Some people regard the advent of off-Earth resource extraction or human settlements as a form of astrocolonialism, which is a term used to describe the extension of colonial practices on Earth to extraterrestrial environments. Astrocolonial critiques have been levied against the emergence of satellite megaconstellations, which are bright enough to interfere with traditional skywatching practices, as well as the prospect of allowing companies to mine outer space bodies for resources, especially if those profits benefit a select few.

In contrast, some thinkers believe that life is such a special phenomenon it should be actively spread to any new environment that could nurture it. This line of reasoning, known as panbiotic ethics, tasks humanity with an almost sacred mandate to ensure that as

Chapter Nine: WHEN WE BECOME THEM ▶ 245

many of Earth's bountiful life-forms can survive beyond its borders for as long as possible.

Supporters of a panbiotic approach to space exploration may defer to philosophical arguments, such as the idea that life's affinity for self-propagation imbues the actual universe with a purpose. To enhance this purpose, humans could pioneer projects to engineer and seed hardy organisms—such as bacteria or rotifers—on extraterrestrial surfaces, in an example of directed panspermia.

"Astroecology shows that life in the galaxy can then achieve an immense future," wrote the chemist Michael Mautner in 2010. "Securing that future for life can give our human existence a cosmic purpose."

▲ A view of a Mars settlement equipped for families. If you have ever had to put a toddler in a snowsuit, you can imagine the challenge of getting one in a spacesuit.

# Future Archaeology

HUMANS HAVE CONTEMPLATED THE PROBABILity and timeline of our own extinction for thousands of years through myths, philosophy, and a perennial expectation of apocalyptic downfall. There's a scientific argument to be made that we are existentially finite; after all, almost everything that ever lived on Earth, from trilobites to tyrannosaurus to our own human cousins, is now extinct. Why should we be the exception to what appears to be a hardline rule of life on Earth?

It's therefore possible, perhaps probable, that humans will never secure a firm foothold off Earth. Perhaps we will go extinct by our own hand, via climate change or nuclear winter, or maybe we'll simply come to be satisfied with our special planetary paradise and abandon our attempts to expand beyond it. Whatever the circumstances, our planet-bound civilization will eventually dwindle and die off, opening up vast niches for new and inventive Earthlings to occupy.

What will be left of our society after our extinction? What might future archaeologists learn from the buried technosignatures of our dead civilization? In his 2008 book *The World Without Us*, which ruminates on this question, journalist Alan Weisman concluded that the most enduring traces of humanity will be bronzeworks, plastic pollution, radioactive waste, monuments like Mount Rushmore, and the fossilized remains of our bodies. If an intelligent alien species arrives

on Earth millions of years in the future—or if another Earthborn civilization arises again—they might be able to piece together a few scant clues about our curious lives and activities from these sources.

Ultimately, though, our longest-lived technosignatures are the probes and messages we've sent to traverse the galaxy far beyond Earth. We've launched five spacecraft so far to wander eternally among the stars, and we've transmitted roughly a dozen radio messages that will continue to sweep outward into space, weakening over time, for eons. Even if we never personally make it beyond our quiet little star system, we've signed our names into cosmic space and time. And whether we ultimately make that first contact or live and die in isolation, we have made our best effort to let others know that, at least, *they* are not alone.

▲ What traces of humanity would be left behind in the event of our extinction?

## AFTERWORD
# Terrestrial Intelligence

I LOVE ALIENS. I HOPE THAT I'VE MADE THAT CLEAR in this book. But I also love humans. Mostly, anyway. I get it: We are dumb. Like, a lot. History is littered with our own goals. We can also be huge jerks, which is perhaps the understatement of the Holocene age.

But hear me out in my closing defense of humans, the only known Earthlings that can (so far) act as conscious ambassadors of our world. As a science reporter, I am privileged to see our species through a rosy lens. Science, a collective enterprise across generations, has allowed us to peer at the dawn of space and time, parse the elemental building blocks of nature, capture waves in the fabric of reality, eradicate diseases that once caused untold suffering, and decode the language of life. My daily grind is chatting with people who roll out of bed to split atoms, launch rockets, and glimpse lost worlds hidden in rock and ice. We are thrillingly clever when we want to be. Our capacity to solve fundamental mysteries with fancy tools is worthy of species-wide pride.

This is not to imply that science makes people better or different than those in any other profession; you can be the best theoretical physicist or biomedical engineer on Earth and still be a blockhead who cannot operate a TV remote. And as much as science inspires wonder,

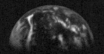

it can—and has, and will continue to be—used for nightmarish ends. The duality of humankind, and all that jazz.

But the routine performance of miracles in the scientific sphere never gets old. That impulse to know things—concretely, empirically, reproducibly—is not only the driving force behind our search for alien life, but also, to my mind, the best justification for it.

As we've seen, many of our tales about aliens imagine them bearing superior technologies or mind-boggling insights from which we could benefit. But since there's no certainty about this outcome, our focus should be as much on what we could offer to another civilization, or even to a simple microbe. They say the best way to find a good partner or circle of friends is to work on yourself first. We have some room for improvement on that front. Our search for others should be animated at least as much by what we have to share as what we stand to gain.

Carl Sagan, who did so much to advance the search for extraterrestrial life, was also one of humanity's best critics. At our finest, he said we are "a way for the cosmos to know itself." Are we the only ones in this colossal universe who have filled in a piece of the ultimate puzzle? In the end, I'm surprised to always find myself more of a Mulder than a Scully on this question. I want to believe—not just in aliens, but in humans, too.

# SOURCES

## CHAPTER 1

Al-Haytham, Ibn, quoted in Tahiri, Hassan. "The Birth of Scientific Controversies The Dynamics of the Arabic Tradition and Its Impact on the Development of Science: Ibn al-Haytham's Challenge of Ptolemy's Almagest." In *The Unity of Science in the Arabic Tradition: Science, Logic, Epistemology, and Their Interactions*, edited by Shahid Rahman, Tony Street, and Hassan Tahiri. Springer (2008). https://doi.org/10.1007/978-1-4020-8405-8_7

Carus, Titus Lucretius, quoted in Wilford, John Noble. "VISIONS: COME OUT, COME OUT, WHEREVER YOU ARE; Get Set to Say Hi to the Neighbors." *The New York Times*, January 1, 2000. https://www.nytimes.com/2000/01/01/news/visions-come-out-come-out-wherever-you-are-get-set-to-say-hi-to-the-neighbors.html

Clarke, P. A. "Australian Aboriginal Astronomy and Cosmology." In *Handbook of Archaeoastronomy and Ethnoastronomy*, edited by Clive L. N. Ruggles. Springer (2015). https://doi.org/10.1007/978-1-4614-6141-8_240

De Fontenelle, Bernard Le Bovier. *Conversations on the Plurality of Worlds*. (1686). Translated by H. A. Hargreaves. University of California Press (1990).

Galilei, Galileo. *The Starry Messenger (Sidereus Nuncius)*. (1610). Translated by The University of Chicago (1989).

Goodman, Matthew. *The Sun and the Moon: The Remarkable True Account of Hoaxers, Showmen, Dueling Journalists, and Lunar Man-Bats in Nineteenth-Century New York*. Basic Books (2008).

Huygens, Christiaan. *Cosmotheoros: Celestial Worlds Discover'd: or, Conjectures Concerning the Inhabitants, Plants and Productions of the Worlds in the Planets, Second Edition*. 1722. Retrieved from Internet Archive. https://archive.org/details/bim_eighteenth-century_kosmotheoros-english-_huygens-christiaan-van_1722

Johnson, Falen. "'We Come from the Stars': Indigenous Astronomy, Astronauts, and Star Stories." *Unreserved*, CBC Radio, January 10, 2021. https://www.cbc.ca/radio/unreserved/we-come-from-the-stars-indigenous-astronomy-astronauts-and-star-stories-1.5861762

Krupp, E. C. *Echoes of the Ancient Skies: The Astronomy of Lost Civilizations*. Harper & Row Publishers, New York (1983).

Locke, Richard Adams. *The Moon Hoax; or, A Discovery That the Moon Has a Vast Population of Human Beings.* The Sun. (1835). Retrieved from Project Gutenberg. https://www.gutenberg.org/files/62779/62779-h/62779-h.htm

Lucian of Samasota. *True History: Introduction, Text, Translation, and Commentary*, edited by Diskin Clay and James Brusuelas. Oxford University Press (2021).

Mu, Tang quoted in Gribbin, J. "Cosmology: The Expanding Universe." In *Galaxy Formation*. Palgrave, London (1976). https://doi.org/10.1007/978-1-349-15657-3_2

Norris, R. P., and B. R. M. Norris. "Why Are There Seven Sisters?" In *Advancing Cultural Astronomy*, edited by Efrosyni Boutsikas, Stephen C. McCluskey, and John Steele. Springer (2021). https://doi.org/10.1007/978-3-030-64606-6_11

Rittenhouse, David, quoted in *Essays, Literary, Moral & Philosophical by Benjamin Rush, M.D. and Professor of the Institutes of Medicine and Clinical Practice in the University of Pennsylvania*. 1798. Retrieved from University of Michigan Library Digital Collections. https://name.umdl.umich.edu/N25938.0001.001.

Roush, Wade. *Extraterrestrials*. MIT University Press (2020).

Rowland, Ingrid. *Giordano Bruno: Philosopher/Heretic.* Farrar, Straus and Giroux (2008).

Schnaufer, Jeff. "Celestial Love Stories." *The Los Angeles Times*, April 1, 1994. https://www.latimes.com/archives/la-xpm-1994-04-01-va-40887-story.html

"The Tale of the Bamboo Cutter" (1600). Retrieved from the Library of Congress. https://www.loc.gov/item/2021667427/

Warmflash, David. "An Ancient Greek Philosopher Was Exiled for Claiming the Moon Was a Rock, Not a God." *Smithsonian*, June 20, 2019. https://www.smithsonianmag.com/science-nature/ancient-greek-philosopher-was-exiled-claiming-moon-was-rock-not-god-180972447/

Weintraub, David. *Religions and Extraterrestrial Life*. Springer Praxis Books (2014).

Whewell, William. *Of the Plurality of Worlds.* 1853. Retrieved from Project Gutenberg. https://name.umdl.umich.edu/N25938.0001.001.

## CHAPTER 2

Bent, Silas. "Mars Invites Mankind To Reveal His Secret." *The New York Times*, August 17, 1924. https://www.nytimes.com/1924/08/17/archives/mars-invites-mankind-to-reveal-his-secret-next-saturday-planet-will.html

Cocconi, Guiseppe, and Philip Morrison. "Searching for Interstellar Communications." *Nature* (1959). https://doi.org/10.1038/184844a0

"Hello, Earth! Hello! Marconi Believes He Is Receiving Signals from the Planets." *The Tomahawk*, March 18, 1920. Retrieved from the Library of Congress. https://chroniclingamerica.loc.gov/lccn/sn89064695/1920-03-18/ed-1/seq-6/

Huang, Su-Shu. "The Problem of Life in the Universe and the Mode of Star Formation." *Publications of the Astronomical Society of the Pacific*, 1959. https://iopscience.iop.org/article/10.1086/127417

Kardashev, Nikolai. "Transmission of Information by Extraterrestrial Civilizations." *Soviet Astronomy*, 1964.

Kellermann, Kenneth, Ellen Bouton, and Sierra Brandt. *Open Skies: The National Radio Astronomy Observatory and Its Impact on US Radio Astronomy*. Springer (2020).

Oberhaus, Daniel. *Extraterrestrial Languages*. MIT University Press (2019).

"Questions to Mars Britons Would Ask." *The New York Times*, January 20, 1929.

"Seeks Sign From Mars In 38-Foot Radio Film." *The New York Times*, August 28, 1924. https://www.nytimes.com/1924/08/28/archives/seeks-sign-from-mars-in-38foot-radio-film-dr-todd-will-study.html

## CHAPTER 3

Adams, Douglas. *The Hitchhiker's Guide to the Galaxy*. Pan Macmillan (2020).
Burroughs, Edgar Rice. *The Chessmen of Mars*. A. C. McClurg (1922).
Chiang, Ted. *Story of Your Life and Others*. Vintage (2016).
Glut, Donald. *Spawn*. Laser Books (1976).
Lem, Stanislaw. *Solaris*. MON, Walker (1961).
Martin, George R. R. *Tuf Voyaging*. Bantam (2013).
McCaffrey, Anne. *Dinosaur Planet Survivors*. Del Ray Books (1984).
Pope, Gustavus W. *Journey to Venus*. Arena Publishing Company (1895).
Ruppersburg, H. "The Alien Messiah in Recent Science Fiction Films." *Journal of Popular Film and Television*, 1987. https://doi.org/10.1080/01956051.1987.9944222
Sagan, Carl. *Contact*. Simon and Schuster (1985).
Sakurazaka, Hiroshi. *All You Need Is Kill*. VIZ Media LLC (illustrated edition) (2014).
Debus, Allen A. *Dinosaurs in Fantastic Fiction: A Thematic Survey*. McFarland & Company (2013).
Silverberg, Robert. "Our Lady of the Sauropods." *Omni Magazine*, September 1980.
Wells, H. G. *The War of the Worlds*. Heinemann (1898).

## CHAPTER 4

"AAF Is Embarrassed; Its Flying Disc Is But a Weather Balloon." *Washington Evening Star*, July 9, 1947. https://www.newspapers.com/article/evening-star/149543672/

Berlitz, Charles, and William Moore. *The Roswell Incident*. Grosset & Dunlap (1980).

Bowman, William. *The Abduction of Betty and Barney Hill: Alien Encounters, Civil Rights, and the New Age in America*. Yale University Press (2023).

Broad, William J. "Urge to Investigate and Believe Sparks New Interest in UFOs." *The New York Times*, June 16, 1987. https://www.nytimes.com/1987/06/16/science/urge-to-investigate-and-believe-sparks-new-interest-in-ufo-s.html

Broad, William J. "Wreckage in the Desert Was Odd but Not Alien." *The New York Times*, September 18, 1994. https://www.nytimes.com/1994/09/18/us/wreckage-in-the-desert-was-odd-but-not-alien.html

Cahn, J. P. "Flying Saucer Swindlers." *True Magazine*, 1952.
Cooper, Helene, Leslie Kean, and Ralph Blumenthal. "2 Navy Airmen and an Object That 'Accelerated Like Nothing I've Ever Seen.'" *The New York Times*, December 16, 2017. https://www.nytimes.com/2017/12/16/us/politics/unidentified-flying-object-navy.html
Frank, Adam. *The Little Book of Aliens*. Harper (2023).
Gaster, Patricia. "'A Celestial Visitor' Revisited: A Nebraska Newspaper Hoax from 1884," *Nebraska History*, 2013. https://history.nebraska.gov/flashback-friday-a-celestial-visitor-revisited-a-nebraska-newspaper-hoax-from-1884-patricia-c-gaster/
Han, Go. "The Possibility of Alien Life Forms and Unidentified Aerial Phenomena." Preprint (2020). https://www.researchgate.net/publication/341822019_The_Possibility_of_Alien_Life_Forms_and_Unidentified_Aerial_Phenomena
Russo, Chris, and Joe Rudy. "How We Staged the Morristown UFO Hoax." *Skeptic Magazine*, April 2009.
Scoles, Sarah. *They Are Already Here: UFO Culture and Why We See Saucers*. Pegasus (2020).

## CHAPTER 5

Greaves, Jane S., et al. "Phosphine Gas in the Cloud Decks of Venus." *Nature Astronomy*, 2021. https://www.nature.com/articles/s41550-020-1174-4
Hand, Kevin. *Alien Oceans: The Search for Life in the Depths of Space*. Princeton University Press (2020).
Johnson, Sarah Stewart. *The Sirens of Mars: Searching for Life On Another World*. Crown (2020).

## CHAPTER 6

Heller, René, and John Armstrong. "Superhabitable Worlds." *Astrobiology*, 2014. https://www.liebertpub.com/doi/abs/10.1089/ast.2013.1088
Kaltenegger, Lisa. *Alien Earths: The New Science of Planet Hunting in the Cosmos*. St. Martin's Press (2024).
Trefil, James, and Michael Summers. *Imagined Life: A Speculative Scientific Journey among the Exoplanets in Search of Intelligent Aliens, Ice Creatures, and Supergravity Animals*. Smithsonian Books (2019).

## CHAPTER 7

Burns, Ruth H. "Ancient Ties: Indigenous Peoples and the Extraterrestrial." *Atmos Magazine*, August 8, 2023. https://atmos.earth/ancient-ties-indigenous-people-and-the-extraterrestrial/

Denning, Kathryn. "Impossible Predictions of the Unprecedented: Analogy, History, and the Work of Prognostication." In *Astrobiology, History, and Society: Life Beyond Earth and the Impact of Discovery*, edited by Douglas A. Vakoch. Springer (2013). https://doi.org/10.1007/978-3-642-35983-5_16

Determann, Jörg Matthias. *Islam, Science Fiction and Extraterrestrial Life: The Culture of Astrobiology in the Muslim World*. I.B. Tauris (2020).

Persson, Erik. "The Moral Status of Extraterrestrial Life." *Astrobiology*, 2012.

Pooley, Jefferson, and Michael J. Socolow. "The Myth of the War of the Worlds Panic." Slate, October 28, 2013. https://slate.com/culture/2013/10/orson-welles-war-of-the-worlds-panic-myth-the-infamous-radio-broadcast-did-not-cause-a-nationwide-hysteria.html

Pullella, Philip. "Vatican Scientist Says Belief in God and Aliens Is OK." Reuters, May 14, 2008. https://www.reuters.com/article/lifestyle/vatican-scientist-says-belief-in-god-and-aliens-is-ok-idUSL1463646/

Saethre, Eirik. "Close Encounters: UFO Beliefs in a Remote Australian Aboriginal Community." *Journal of the Royal Anthropological Institute*, 2007. https://rai.onlinelibrary.wiley.com/doi/abs/10.1111/j.1467-9655.2007.00463.x

Wickramasinghe, Chandra. "Life in the Universe: Concordance with Buddhist Thought." *The Journal of Oriental Studies*, 2014.

# CHAPTER 8

Behroozi, Peter, and Molly S. Peeples. "On the History and Future of Cosmic Planet Formation." *Monthly Notices of the Royal Astronomical Society*, 2015. https://doi.org/10.1093/mnras/stv1817

Bostrom, Nick. "Where Are They? Why I Hope the Search for Extraterrestrial Life Finds Nothing." *MIT Technology Review*, 2008.

Forgan, Duncan H. "Spatio-temporal Constraints on the Zoo Hypothesis, and the Breakdown of Total Hegemony." *International Journal of Astrobiology*, 2011. https://www.cambridge.org/core/journals/international-journal-of-astrobiology/article/abs/spatiotemporal-constraints-on-the-zoo-hypothesis-and-the-breakdown-of-total-hegemony/06D2915B2AABA203BE0701E644C6C1C9

Hanson, Robin. "The Great Filter: Are We Almost Past It?" September 15, 1998. Retrieved from George Mason University. https://mason.gmu.edu/~rhanson/greatfilter.html

Liu, Cixin. *The Dark Forest*. Tor Books (2016).

Musso, Paolo. "The Problem of Active SETI: An Overview." *Acta Astronautica*, 2012. https://www.sciencedirect.com/science/article/abs/pii/S0094576511003791

Shostak, Seth, quoted in Pethokoukis, James. "Keeping His Eyes on the Skies." *U.S. News & World Report*, November 4, 2003.

Tipler, Frank. "Extraterrestrial Intelligent Beings Do Not Exist." *Royal Astronomical Society, Quarterly Journal*, 1980. https://pubs.aip.org/physicstoday/article-abstract/34/4/9/433414/Extraterrestrial-intelligent-beings-do-not-exist

Ward, Peter, and Donald E. Brownlee. *Rare Earth: Why Complex Life Is Uncommon in the Universe*. Copernicus Publications (2000).

# CHAPTER 9

Lempert, William. "From Interstellar Imperialism to Celestial Wayfinding: Prime Directives and Colonial Time-Knots in SETI." *American Indian Culture and Research Journal*, 2021. https://doi.org/10.17953/aicrj.45.1.lempert

Mautner, Michael N. "Seeding the Universe with Life: Securing Our Cosmological Future." *Journal of Cosmology*, 2010. https://thejournalofcosmology.com/SearchForLife111.html

Noon, Karlie Alinta. "Indigenous Cosmic Caretaking and the Future of Space Exploration." Centre for International Governance Innovation (CIGI), 2024. https://www.cigionline.org/static/documents/Noon-Sept2024_F9BMunh.pdf

O'Neill, Gerald. *The High Frontier: Human Colonies in Space*. William Morrow and Company (1976).

Weisman, Alan. *The World Without Us*. St. Martin's Press (2007).

White, Frank. *The Overview Effect: Space Exploration and Human Evolution*. American Institute of Aeronautics (2014).

# INDEX

Page numbers in *italics* indicate illustrations and their captions.

abductions, 5, *5*, 107–109, 110–111
"Across the Universe," 50
active SETI, 46
Advanced Aerospace Threat Identification Program, 116
Aetherius Society, 115
aircraft, mistaken for UFO, 94
"Alien Messiah, The" (Ruppersburg), 63
aliens, derivation of term, 8
Aliens franchise, 59, *212*
Allan Hills 840001 (meteorite), 129–130
Allen, Paul, 167
Allen Telescope Array, 167
Almár, Iván, 167
Alpha Centauri, 158, 186
Anaxagoras of Clazomenae, 9
Andromeda, 160
animals, 2
Area 51, 111
Arecibo Message, 49, 50
Arecibo Observatory, *219*
Aristarchus of Samos, 15
Aristotle, 12
Armstrong, John, 185
Arnold, Kenneth, 84, 85, 90
Artemis program, 243
artifacts, as biosignatures, 171–172
Asimov, Isaac, 47
astrocolonialism, 244

astronomical units, 158
astrotheology, 12
Atchley, Dana, 43
Atlas, 6
atomism, 11–12

backward contamination, 153
balloons, 94, *94*
Barnard's Star, 158
Beatles, 50
Bent, Silas, 31, 33
*Beresheet*, 153, 214
Berlitz, Charles, 105
Bermuda Triangle, 111
Beserker hypothesis, 226–227, *226*
Beta Pictoris system, 165, *165*
Bialy, Shmuel, 152
Big Ear telescope, 45–46
Bigelow, Robert, 111
Bikini Atoll, 104
biofluorescence, 169, *169*
biosignatures, 38, 128, 157, 167, 168
biospheres, relict, 139
Blob, 59
Borisov, Gennadiy, 152
Bostrom, Nick, 225
Bower, Doug, 99
Brazel, Vernon, 85
Brazel, W. W. "Mac," 85–86
Breakthrough Listen initiative, 167

Breakthrough Starshot project, 186
Breslow, Ronald, 81
Broad, William, 113
*Brookings Report*, 197
Brownlee, Donald E., 220–222
Bruno, Giordano, 18–19, *18*
Buddhism, 209
Byurakan Astrophysical Observatory, 42

Cahn, J. P., 97
Calhoun, J. D., 95
Callisto, 134
Calvin, Melvin, 43
Carpenter, John, 69–70
*Cassini* orbiter, 144
Catholic Church, 17, 18
cattle mutilation, 99
celestial objects, 4–5, 9. *See also individual objects*
Ceres, 134
Chiang, Ted, 75
chiral molecules, 168
CHNOPS formula, 144
Chorley, Dave, 99
Christianity, 12, 62, 207
CIA (Central Intelligence Agency), 114
climate change, 224
Cocconi, Guiseppe, 36–38
Colbert, Stephen, 50
colonialism, 195–196, 244
colonies, off-Earth, 240–242

Comet Borisov, 152
comets, 92
Committee on the Peaceful Uses of Outer Space, 204
communications
 attempts at, 48–51
 potential, 37–38, 45–46
 radio, 30–31, 37, 169–170, 202–203
 search for, 169–170
Condon report, 102
Cook, James, 195
Copernican model, 17, 18
Copernican principle, 16
Copernican revolution, 15–16
Copernicus, Nicolaus, 16, 18
coronagraphs, 165, 174–175
Cortez, Hernán, 195
cosmic pluralism, 8, 12, 18, 21
cosmological horizon, 160
Craigslist, 50
Cronus, 8
crop circles, 98–99, *98–99*
*Curiosity* rover, 130

Dark Forest hypothesis, 225
Darwin, Charles, 222
Davinci, 139
Davis, Joe, 49
*Declaration of Principles Concerning Activities Following the Detection of Extraterrestrial Intelligence*, 203, 204–205
Deep Space Communications Network, 50
Denisovans, 2
Denning, Kathryn, 197
Descartes, René, 21
Determann, Jörg Matthias, 206–207
dinosaurs, 76–81, *77*
Dione, 134

direct imaging, 165, 174–175
*District 9*, 67, *68*
divine madness, 70
divinities, belief in, 4
Doritos, 50
*Dragonfly*, 148, 201, *201*
Drake, Frank, 40–44, 166
Drake equation, 41–42
Dundy County hoax, 95–96
Dyson sphere, 170–171, *171*


Earth, search for origin of life on, 52–53
Earth twins, 182–183, *182*
Earth zoo hypothesis, 229
eclipses, 9, 15, 176
Ehman, Jerry R., 46
Einstein, Albert, 176
Enceladus, 123, 134–135, *135*, 144, *144–145*
Enceladus Life Finder, 144
energy consumption, categorization by, 43–44, 170, 171
Epsilon Eridani, 38, *38–39*, 40, 49
ethics/morals, 210–215, 243–245
Europa, 123, 134–135, *134*, 140–142, *140*, *141*
*Europa Clipper*, 142
European Southern Observatory, 175
exodinosaurs, 76–81
ExoMars mission, 132
exoplanets, 38, 156, 164–165, 182–185, 186–187
exosociology, 205
exotheology, 12
extinction, 77–78, 246–247
extraterrestrial life, *247*
Extremely Large Telescope (ELT), 175, *175*


fata morgana, 93
Federation of American Scientists, 114

Fermi, Enrico, 35–36, 223–224
Fermi paradox, 35–36, 218, 220, 223, 224, 225, 230
fireworks, 94
"first contact" moments, 195–197, 199–203
Firstborn hypothesis, 230–231
Five-hundred-meter Aperture Spherical Telescope (FAST), 167
flap, 96
flares, 94
Flying Saucer Working Party (FWSP), 103
flying saucers, history of term, 90. *See also* unidentified flying objects (UFOs)
folk religions, 209
Fontenelle, Bernard Le Bovier de, 20–22
foo fighters, 89
Forgan, Duncan, 229
forward contamination, 153
forward-looking infrared (FLIRs) targeting instruments, 116
Frank, Adam, 164
Fravor, David, 116
Freudenthal, Hans, 51
Friedman, Stanton, 105
Funes, José Gabriel, 207


Gagarin, Yuri, 236
Galactic Habitable Zone (GHZ), 177
Galileo Galilei, 17, 140
Ganymede, 134
gases, as biosignatures, 168
Gebauer, Leo A., 96–97
generation ships, 241
geocentrism, 12, 18
geopolitics, galactic, 74
Gernhardt, Michael, *237*
Glaser, Hans, 88, *89*
Gliese 445, 234

Gliese 526 system, 51
Gliese 581c system, 50
gravitational fields, 21
gravitational lensing, 176
Great Filter, 223–225, 227
Great Moon Hoax, 24
Green Bank Observatory, 36, *37*, 43
Grusch, David Charles, 117
G-type stars, 180

Habitable Worlds Observatory, 174–175
habitable zone, 38, 177, 180
hallucinations, 94
Hanson, Robin, 223, 224
Hart, Michael, 227
Hart-Tipler Conjecture, 227
Hawking, Stephen, 195–196
heliocentric model, 16, *16*, 17
Heller, René, 185
Helmholtz, Hermann von, 146
Herschel, John, 24
Hill, Barney and Betty, 107–109, *107*, *108*
Hinduism, 12, 208–209
Hipparchus, 3
Howard E. Tatel Radio Telescope, 40
Huang, Su-Shu, 36, 38, 40, 43
Huang, Yingrui "Zao," 215
Hubbard, L. Ron, 115
human-alien bureaucracies, 74
human-alien sex, 10, 65–66
human-alien shapeshifters, 71–73
Huygens, Christiaan, 25
*Huygens* probe, 148
hydrogen line, 37, 40, 45

Ibn al-Haytham, 15
infrastructure, planetary, 170
interior water ocean worlds (IWOWs), 133–134

International Academy of Astronautics (IAA), 204, 205
International Space Station, *239*, 240
International UFO Museum and Research Center, 118–119
interstellar objects, 149–152
Islam, 12, 206–207

James Webb Space Telescope, *162–163*, 167, *181*
Jenkins, Charles Francis, 31–32, *31*
Jezereo Crater, 131, *131*, 133
Judaism, 13
Jupiter, 17, *141*, 142, 158. *See also* Europa

Kaltenegger, Lisa, 81
Kardashev, Nikolai, 43–45, 166, 230
Kardashev scale, 43–44, 170, 171
Kayuga-hime, Princess, 14, *14*
Kepler, Johannes, 17
Kepler telescope, 164
Kepler-442b, 187
King, George, 115
Klaatu, 62, *62*
Konopinski, Emil, 35
Krishna, *208*
K-type stars, 180, 187

Labeled Release (LR) experiment, 128–129
lasers, megawatt-class, 203
lens flare, 94
lenticular clouds, 93, *93*
Lescaze, Zoë, 78
Levin, Gilbert V., 129
light curves, 164
Lilly, John, 43
Lincos, 51, *51*
Liu Chixin, 225

Local Group, 159
Locke, Richard Adams, 24
Loeb, Avi, 152
Lowell, Percival, 33, *33*
Lucian of Samosata, 10, 65
Lunine, Jonathan, 202

MacLachlan, Kyle, 118
Maimonides, 13, *13*
Marcel, Jesse, 104–105, *105*, 112–113, 118
Marconi, Guglielmo, 33
Mars, 30–33, *33*, 46, 123, 126–133, 128–129, *198*, 199–201, 212–213, 241, *242*, *244*, *245*
Mars Sample Return (MSR), 131, 133, 199–200
Martin, Joe, 90
Mass Effect series, 66
Mautner, Michael, 245
messages, search for in sky, 4
Messaging to Extraterrestrial Intelligence (METI), 47
messianic narratives, 62–63
meteorites, 92, 129
meteors, 92
methane, 130
microlensing, 165, 174
microscopy, 23
Milky Way, size of, 159
Millstone Hill Radar, 49
Milner, Yuri, 167
Mimas, 134
molecular chirality, 168
monogatari, 14
monsters, aliens as, 57–59
Moon, 14, 24, *24*, *26*, *34*, 183, 221, *222*, 236, 243
Moore, William, 105
morals/ethics, 210–215, 243–245
Morrison, Herbert, 192
Morrison, Philip, 36–38, 43
Morristown hoax, 97–98

Morse Message, 48
Mount Wilson Observatory, 156
music, aliens in, 80
Musk, Elon, 132
Musso, Paulo, 227
Mutual UFO Network (MUFON), 93

Nancy Roman Space Telescope, 174, *174*
National Radio Astronomy Observatory (NRAO), 40
Neanderthals, 2
Newton, Silas M., 96–97
Nicholas of Cusa, 18
Norris, Ray and Barnaby, 6

Oliver, Barney, 43
O'Neill, Gerard K., 241
O'Neill cylinders, 240–241, *240*
Operation Crossroads, 104
*Opportunity* rover, *198*
opposition, of planets, 31
optical beacons, 202–203
orange dwarfs, 180, 187
"Order of the Dolphin," 40–41, 43
Orion, 6
'Oumuamua, 149–152, *150–151*, 172

panbiotic ethics, 244–245
panspermia, 146, 201, 245
pantropy, 236–237
Passaic UFO photos, *90–91*
Pearman, J. P. T., 43
peers, aliens as, 64–66
Pentagon UFO videos, 116
Pericles, 9
*Perseverance*, 131, 133
Persson, Erik, 214
phospine, 138
photosynthetic life-forms, 52, *53*
Pike, John E., 114
*Pioneer* plaques, 48, *48*

planetary chauvinism, 242
planets
  opposition and, 31
  total number of, 42
  *See also individual planets*
Plato, 12
Pleiades star cluster, 6–8, *6–7*
Pleione, 6
Pluto, 134, 143, 158
Polaris, 50, 51
pollution, as biosignature, 170
possession, 5
post-detection scenarios, 199–215
Prime Directive, 211–212, 229
"Problem of Life in the Universe and the Mode of Star Formation, The" (Huang), 36, 38
Project Blue Book, 102
Project Cyclops, 44
Project Grudge, 102
Project Magnet, 103
Project Mogul, 112–113, *113*
Project Ozma, 40, 42
Project Sign, *100–101*
Prometheus, 4
Proxima b, 186, *186*
Proxima Centauri, *159*
Ptolemy, Claudius, 15

radio communications. *See under* communications
radio photo message continuous transmission machine, 31
radio telescopes, 175
Raëlism, 115
Ramey, Roger, 86
Readick, Frank, 192
red dwarfs, 178–179, *179*, 184, 231
Reeve, Christopher, 60
relativity, general theory of, 176
relict biospheres, 139

religion, 11–13, 17, 18, 62, 115, 206–209
Rio scale, 167
Rittenhouse, David, 25
Rommel, Kenneth, 99
*Rosalind Franklin*, 132
Roswell, New Mexico, 84–86, 104–106, 112–114, 118–119, *119*
Roswell Army Air Field (RAAF), 86
Roswell Report, *114*
Rudy, Joe, 97–98
Rukmini, *208*
Ruppersburg, Hugh, 63
Russo, Chris, 97–98

Sagan, Carl, 5, 43, 72, 166, 211
satellites, 94
saviors, aliens as, 60–63
science fiction, 10, 34–35, 59
scientific revolution, 23–25
scientific revolution, aliens and, 23–25
Scientology, 115
Scully, Frank, 96, *96*
"Searching for Interstellar Communications" (Cocconi and Morrison), 36–38
searchlights, 94
seasons, 168–169
SETI Permanent Committee, 205
SETI research
  beyond Milky Way, 189
  biosignatures sought by, 166–172
  developments in, 166–167
Seven Sisters story, 6, 8
Shkadov thrusters, 171
Shostak, Seth, 230
Shuster, Joe, 60
Siegal, Jerry, 60
Simple Response to an Elemental Message, A (ASREM), 51

Skinwalker Ranch, 111
sky lanterns, 94
Smith, William Brockhouse, 103
Solar Gravitational Lens (SGL), 176–177
solar system
 image of, *123*
 knowledge of, 122–123
 size of, 158
SpaceX, 132
spectroscopy, 164
Square Kilometer Array (SKA), 175
Star Trek franchise, 64, 65, *65*, 71, 74, 81, 211, 229
Star Wars franchise, 64–65, 74
stars and star systems, types of, 178–181
sterilization of spacecraft, 214–215
Stock, George, *90–91*
strangers, aliens as, 69–70
subordinates, aliens as, 67–68
sun dogs, 93
Super-Earths, 184, 187
superhabitable worlds, 184–185
Superman, 60–62, *61*
supernovas, 181

Tabby's Star, 170–171
"Tale of the Bamboo Cutter, The," 14
tardigrades, 153, *153*, 214
Tarter, Jill, 166, 167
Tau Ceti, 38, 40, 49
technologies, invisible, 230
technosignatures, 38, 157, 166, 175, 189, 246–247
telescopes, advances in, *32*, 34, 36, 174–177
Teller, Edward, 35
temporal habitable zone, 143
Tenchtitlan, *196*
terminator habitable, 184

terraformed worlds, 241, 242
Tesla, Nikola, 32
tidally locked planets, 179, 183–184, *183*, 187
time-distorting aliens, 74–75
Tipler, Frank, 227
Titan, 123, 134, 143, 147–148, *147*, *148*, 201–202
Titania, 134
Todd, David Peck, 31–32, *31*
transit method, 164
Transiting Exoplanet Survey Satellite (TESS), 164
TRAPPIST-1 system, 187
tropes, of aliens, 71–75
Turyshev, Slava, 177
21-centimeter signal, 37

UFO Festival, 118–119
ufology, origins of, 84
ultracool dwarfs, 187
unidentified anomalous phenomena (UAP), 84, 116–117
unidentified flying objects (UFOs)
 explanations for, 92–94
 government reactions to, 101–103, 113–114
 hoaxes involving, 94, 95–99
 increase in reports of, 35
 at Nuremberg, 88, *89*
 photos of, *90–91*, *117*
 Roswell and, 85–86
 during Song dynasty, 87–88
universe, size of, 158–161, *162–163*

Van Maanen's Star, 157
Vatican Observatory, 207, *207*
Venus, 48, 78–79, 92, 123, 127, 135–139, *137*, *138*, 241

Venus Life Finder, 139
Vera C. Rubin Observatory, 152
Veritas, 139
*Viking* landers, 128–129, 212–213
Villas Boas, Antônio, 110
visitations, 5, 14
von Neumann, John, 227
Vorilhon, Claude, 115
*Voyager* probes, 49, *49*, 140, 176, 226, 234, *235*

Walton, Travis, 110
*War of the Worlds, The* (Wells), 192–193, *193*
Ward, Peter, 220–222
Watterson, Bill, 76
Weisman, Alan, 246
Welles, Orson, 192–193, *193*
Wells, H. G., 57, 59, 192
Whewell, William, 27, *27*
White, Frank, 237
white dwarfs, 181
Wickramasinghe, Chandra, 209
Wolf-Rayet star, *181*
Wolski, Jan, 111
womb worlds, 133–134
*World Without Us, The* (Weisman), 246
Wow! signal, 45–46, *45*, 50

yellow dwarfs, 180
Yevpatoria Planetary Radar complex, 48, 49, 50
York, Herbert, 35

Zeus, 5
zoo hypothesis, 228–229, *229*

# ACKNOWLEDGMENTS

This book was born just before the winter holidays of 2022, when my extraordinary colleague Samantha Cole connected me with extraordinary editor John Meils. I am so grateful to them both for helping me achieve my lifelong dream of writing a raucous book about a topic I adore. I feel so lucky that John was my editorial guide through this process; his talent, humor, wisdom, and patience were a constant source of comfort and inspiration.

I'm indebted to Sarah Levitt, my agent at Aevitas Creative, who believed I could write a book well before I did. Her boundless passion and fierce advocacy encouraged me every step of the way. Thanks also to Danny Cooper, who stewarded the book to its finish and made it so much better, as well as the entire Workman Publishing team for making it possible.

I spent more than a decade covering space, aliens, and other weird stories for Motherboard, a scrappy site that punched way above its weight. I owe my career to the guidance and creativity of my editors there, especially Jordan Pearson (o7), Emanuel Maiberg, Brian Merchant, and Jason Koebler. I feel fortunate to have friends who have offered me support, advice, commiseration, and (most importantly) cake or candy throughout this process, and so many others. In particular, I'd like to shout out Chris Dingwall, Josh Bowman, Simon McNabb, Jen Neale, Laura Blue, Genevieve Szyablya, Carla Jeanpierre, Alexandra Blacker, Maddie Stone, Kate Lunau, Dan Hershfield, Nicole Drespel, Rachel Bloom, and Peter Gerena.

My interest in science was sparked from a young age by my parents, Sue and Paddy, who expressed an enthusiasm for learning in their careers as doctors and at home, where "life, the universe, and everything" was a topic of frequent rumination. Mum, you are my North Star and my best friend. Dad, you are one in a trillion. Thanks for taking such good care of me, and Janice, thanks for taking such good care of Dad.

Ben, big brother, I remain in awe of your talent and drive. Lucy, if everyone had a sister like you, there would be world peace. You are quietly incredible. Ian, you are the epitome of a great bloke. And Rebecca and Rhys, Auntie Becky adores you and can't wait to hunt some dragons again soon. Thanks also to my in-laws, Steve and Robin, for all your help, generosity, and impromptu haircuts.

Kyle, you are my habitable world. I'm glad you found me. And Luke, I can't believe that in this vast universe, filled with infinite possibilities, I got to be your Mommy. Total score. Whether or not we find life on other worlds, I am so proud to share my life here on Earth with you and Daddy.

# PHOTO CREDITS

**COVER: Shutterstock:** FeelplusCreator

**Alamy:** AJ Pics p. 70; Allstar Picture Library Limited p. 65; Allstar Picture Library Ltd p. 73; Fabrizio Annovi p. 47; Cavan Images p. 173; Charles Walker Collection p. 108; Chronicle p. 27; Collection Christophel p. 212; CPA Media Pte Ltd p. 14; Elen p. 182; Entertainment Pictures p. 68; Glasshouse Images p. 32; LoveEmployee p. 171; NASA Image Collection p. 145; Photo 12 p. 193; Pictorial Press Ltd pp. 61, 62, 63; Zev Radovan p. 13; Vladimir Razguliaev p. 94; RealyEasyStar/Fotografia Felici p. 207; Science Photo Library p. 143; Stocktrek Images, Inc. p. 39; Marcel Strelow p. 169; Colin Waters p. 85; ZUMA Press, Inc. p. 119. **AP Images:** NASA/John Hopkins/APL p. 201. **Courtesy Use:** Carnegie Institution for Science p. 157; Department of Defense, HO / NYT p. 117; ESO Images/M. Kornmesser pp. 150–151; North American AstroPhysical Observatory (NAAPO) p. 45. **Getty Images:** Archive Photos p. 66; Bettmann p. 107; duncan1980 p. 26; MediaProduction p. 5. **NASA:** Pablo Carlos Budassi p. 183; ESA/Hubble & NASA, J. Lee p. 188; ESA/NASA/JPL/University of Arizona p. 148; ESO/M. Kornmesser p. 186; NASA pp. 134–135. 162–163, 238–239; NASA/WMAP Science Team p. 161; NASA, ESA, CSA, STScI, Webb ERO Production Team p. 181; NASA/Chris Gunn p. 215; NASA/JPL-Caltech pp. 123, 133, 127, 131, 179, 124–125; NASA/JPL-Caltech/SSI/Kevin M. Gill p. 147; NASA/JPL/Cornell p. 198; NOAA p. 221; S. Turyshev p. 177. **National Radio Astronomy Observatory:** Jeff Hellerman, NRAO/AUI/NSF p. 166. Science Source: Richard Bizley pp. 53, 226; Detlev Van Ravenswaay p. 222; Sebastian Kaulitzki p. 228; KTSDesign p. 110; Library of Congress p. 31; Magrath/Folsom p. 93; NASA p. 34; NASA Ames Research Center/Don Davis p. 240; New York Public Library pp. 33, 196; Dr. Seth Shostak p. 37; Steven Hobbs/Stocktrek Images pp. 244–245; Phil Wilson/Stocktrek Images p. 77. **Shutterstock:** DFLC Prints p. 18; Vladimir Mulder p. 247. **The European Space Agency:** ESA/Juice/JANUS p. 249; **The Public Domain Review:** Duke University Libraries p. 58.

**Wikimedia: Public Domain:** Aus fernen Welten by Bruno H. Brugel p. 3; Aubrey Beardsley p. 10; Andreas Cellarius p. 16; ESO/A.-M. Lagrange et al. p. 165; HannahMoss p. 23; Jabberocky p. 99; Library of Congress p. 24; Los Angeles County Museum of Art p. 208; NASA pp. 48, 174, 237; NASA/JPL/DLR pp. 17, 140, 235; National Space Science Data Center p. 49; Records of Headquarters U.S. Air Force p. 103; Science Source p. 21; George Stock p. 91; The Soviet Space Program p. 138; United States Air Force pp. 86, 100, 113; Zentralbibliothek Zürich p. 89.

**3.0 License:** Daein Ballard p. 242; Skatebiker p. 159.

**4.0 License:** European Southern Observatory p. 175; Feoffer p. 96; Fort Worth Star-Telegram p. 105; NASA/JPL-Caltech/SSI/CICLOPS/Kevin M. Gill p. 141; Nuremberg_chronicles p. 9; Powellelli p. 115; Tedder p. 219.

**2.0 License:** JAXA/ISAS/DARTS/Kevin M. Gill p. 137; M45 Pleiades p. 7.

**2.5 License:** Schokraie E, Warnken U, Hotz-Wagenblatt A, Grohme MA, Hengherr S, et al. (2012) p. 153.

# ABOUT THE AUTHOR

Becky Ferreira is a science reporter interested in space, dinosaurs, ancient cultures, cool rocks, and lying motionless on couches. She is a contributor to the *New York Times*, *WIRED*, and *National Geographic*, among others, and writes the Abstract column for 404 Media. She lives with humans and felines in Ithaca, New York.

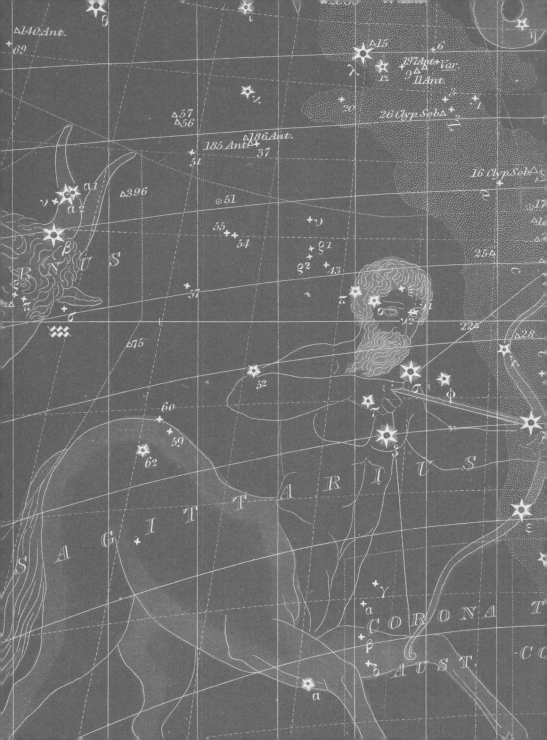